PRAISE FOR *TRANSFORMING THE UTILITY POLE*

Not all disruptive innovations are made in Silicon Valley or from bytes and bits. This one comes from South Carolina's wood waste and reveals hidden value just waiting for utility companies to capture. Thanks to Breede's direct experience, Transforming the Utility Pole *is an easy, compelling read. This book is an example of authentic, evergreen, entrepreneurship—creating value for others from an ignored resource—for both the common good and economic benefit. This is real innovation and real value.*

—LANNY VINCENT

General Partner, Vincent & Associates, Ltd.

As the implementation and management of corporate sustainability initiatives become increasingly important to utilities, they are placing greater value on suppliers that can develop innovative products and services to help them manage their waste streams. This book provides an interesting look at how companies, even those operating in commodity businesses like poles, can differentiate themselves by providing innovative, unique, and cost-effective solutions for their customers

—CATHERINE HEIGEL

Past President, Duke Energy South Carolina

Circular business models are not for every business, but every business must seriously evaluate and consider what it would take to turn liabilities or waste into assets. It's too easy for academics and consultants to allude the value of circular economics; it's quite different to learn from those that have, and are, perfecting such models. Kudos to Barry's dose of reality!

—JOHN MEINDL

Executive-in-Residence, Furman University

Corporate Strategy and Sustainability

As utilities nationwide continue to look for more answers on how to properly dispose of used utility poles, the goal of achieving a balance between overall cost and environmental sustainability becomes increasingly important. This book does an excellent job of both pointing out the challenges related to disposal while also offering bona fide solutions. Well done!

—ARCHIE LOPEZ

Director of Strategic Initiatives, Texas Electric Cooperatives

TRANSFORMING
THE UTILITY POLE

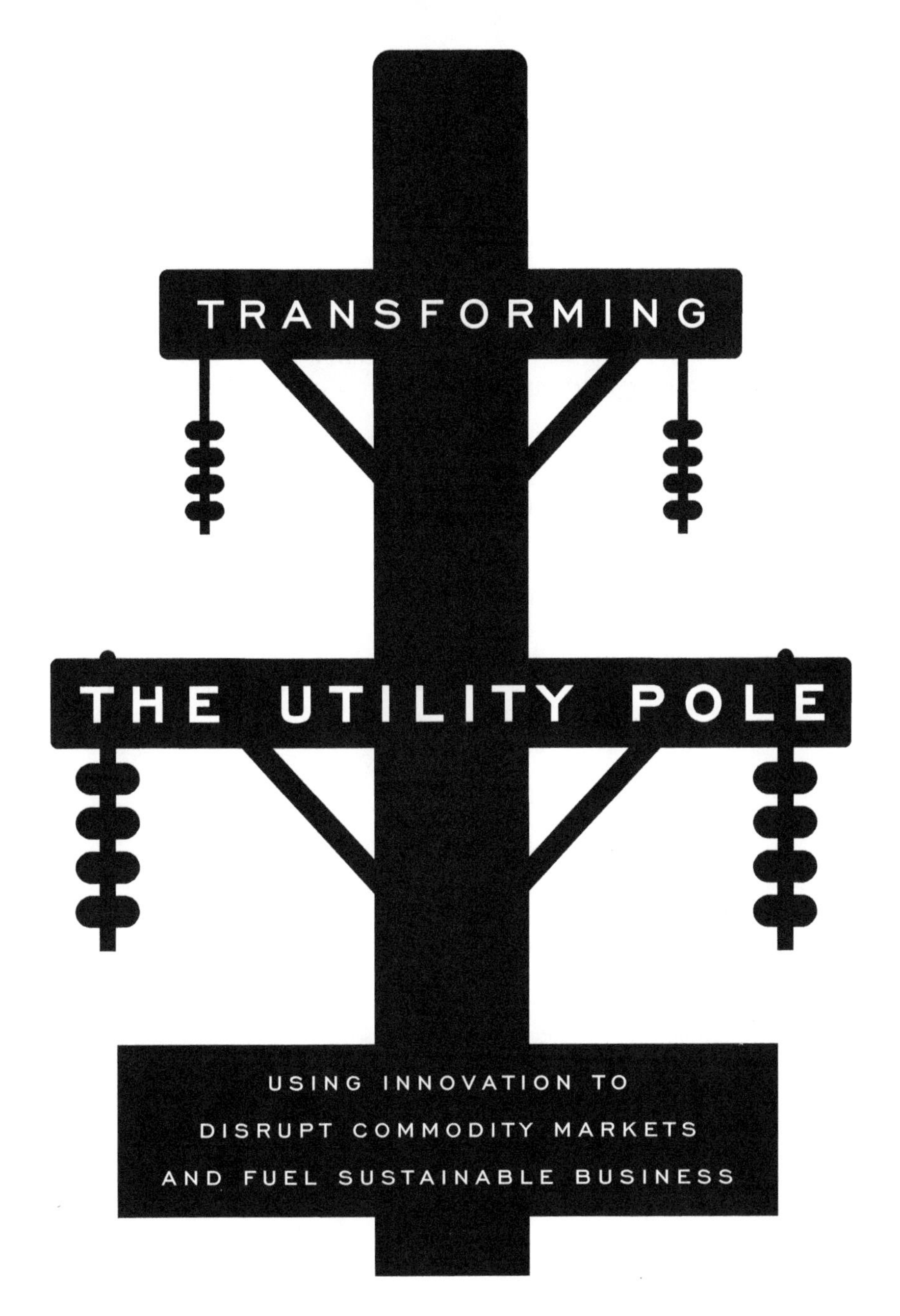

BARRY BREEDE

Advantage®

Published by Advantage, Charleston, South Carolina.
Member of Advantage Media Group.

Printed in the United States of America.

10 9 8 7 6 5 4 3 2 1

ISBN: 978-1-59932-795-2
LCCN: 2018953903

Cover design by George Stevens.
Layout design by Megan Elger.

TABLE OF CONTENTS

ACKNOWLEDGEMENTS

Inspiration for this book arrived from many different people and places—too many to list in total. A special thanks to the management, board, and family ownership of Cox Industries and their steadfast belief in the value of innovation as a driver of future business success. Additional thanks to the many terrific people in the utility industry that I've had the pleasure of working with, especially those who continue to carry the torch for building more sustainably focused businesses. Thanks also to the people of Koppers whose investment in Cox vividly illustrates their parallel interest in the continued pursuit of innovation. Lastly, a special thanks to Nancy, Jennifer, Mary Anne, and Robert. They are four individuals who form one family that provides much-needed love and support.

A NOTE FROM THE AUTHOR

During the writing of this book, Cox Industries was acquired by Koppers and renamed Koppers Utility and Industrial Products, thus ending the Cox family's sixty-plus-year ownership of the business. Koppers is a global provider of wood treatment chemicals, and is the largest producer of treated wood railroad ties in North America. With the addition of Cox, Koppers is poised to become one of the world's largest wood treatment, production, and supply companies.

Rather than changing the substance of the lessons drawn in this book about mid-sized, family-owned businesses, this acquisition presents an opportunity for Cox as an innovator to take the next step in our journey, allowing us to marry our thinking about wood utility poles with the ongoing work that Koppers is doing in wood rail ties. In particular, our shared interest in developing environmentally sound disposal methods for these products should help accelerate our innovations in this area and fuel sustainable business.

In the following pages, I discuss a major innovation, pioneered at Cox, in dealing with used utility poles made of chemically treated wood. There is no reason, of course, that the process we have introduced should be restricted to utility poles. The chemical treatment process for railroad ties is much the same, and we now have the opportunity to apply our innovation thinking to an even larger source of wood waste—enhancing both our economic and environmental impact.

—BARRY BREEDE, MAY 2018

PREFACE

This book is about utility poles, but actually it is about a lot more than that.

First, it is about a South Carolina company, Cox Industries, that I hope a lot of leaders in small to mid-sized businesses will see themselves reflected in. Since 1954, Cox has been a leading manufacturer and distributor of high-quality treated wood products for the residential, commercial, industrial, and utility markets. Our core product is the wood of the southern yellow pine tree, which we harvest, cut into pieces of all shapes and sizes, treat with a chemical to extend its life span, and sell in the form of everything from telephone poles to fence posts and porch spindles.

Family owned and operated since day one,[1] Cox has grown into one of the largest providers of treated wood utility poles in the United States. It has evolved from a two-person leadership structure to a diverse board of directors with expertise in several areas. In this way, it has followed the trajectory of a lot of companies that have seen impressive growth in the last several decades. Still, it remains tight-

1 Acquired by Koppers in 2018.

knit and family operated; the CEO, Mikee Johnson, is the grandson of Bill Cox Sr., who cofounded the company with his brother.

Second, this book is about innovation. As I explain, Cox was founded on the basis of an innovation in the treatment of wood, and it remains well placed to innovate in this industry to this day. Unsatisfied with a random or one-off approach to generating new ideas, however, it has implemented a change-oriented model, a full-time innovation staff, and a sequential innovation process that can serve as an exemplar to companies across industries that want to implement profitable innovations and market-changing disruptions more effectively.

Finally, it is about sustainability. Cox maintains a Sustainability Advisory Panel made up of university professors, philanthropists, conservationists, and business leaders who together provide a variety of perspectives on environmental issues, society, and the bottom line. Wood is an excellent product that provides many years of useful life, but it presents a unique challenge at the end of that life: how to dispose of it. The social and environmental costs of current practices, especially among utilities that are disposing of their used poles, are too high to maintain. With landfills reaching capacity, we in the industry have no choice but to seek alternative disposal methods; otherwise, we are on a collision course with disaster.

Through its targeted sequential innovation process, Cox came to focus on this problem, and has since become a leader in environmentally sensitive disposal practices. The story of this book is the story of that development, transforming the utility pole both literally and figuratively, and in the process moving into entirely new business and developing entirely new products. And it is the story of the power of innovation itself, and how it can be harnessed to disrupt commodity markets and fuel sustainable business.

Here again, I hope the example of Cox and its innovations in the environmental realm serve as an inspiration to business leaders across industries who are interested in learning how to make sustainability and corporate responsibility work together with running a successful and profitable business.

CHAPTER 1

The Cox Tradition of Innovation

Say that you want to lower the transportation costs for shipping your product. A clear solution would be to ship more of the product per shipment, but here are the conditions: The product has to go on a truck, and that truck can only hold so much weight. Each individual unit of your product has a certain weight, and the trucks are maxed out.

Dead end?

Not if you can make the product weigh less. That's an opportunity for innovation.

When people think of companies that launch and grow themselves on the basis of innovation, they tend think of a handful of choice industries or sectors: information technology (IT), automobiles, pharmaceuticals, and so forth. The forest products industry probably does not immediately come to mind. Still, Cox Industries has a reputation for advancing the industry and changing the market in the forest products sector. Its original innovation, incidentally, developed in response to the problem of saving money on shipping its product.

Cox is a family-owned business, founded in the 1950s by brothers W.B. and E.J. Cox, that manufactures and distributes treated wood products ranging from lumber for residential building to poles for purchase and use by utilities. W.B. Cox—Bill Sr., grandfather of current Cox CEO Mikee Johnson—was driven from the beginning to keep coming up with new ways of making his company more efficient and profitable. So when faced with the limitations on how much wood he could get onto a truck, he innovated.

When wood is chemically treated, it is soaked with a liquid carrier that helps the chemical penetrate into the wood, so the wood comes out of the process soaking wet. Traditional practice had always been to let the wood air-dry over the course of several days; during that time, it would be shipped wherever it needed to go. The wood, of course, was much heavier when wet, which limited the amount that could be loaded onto a truck for the initial shipping. Bill Cox Sr. came up with a method—kiln-drying after treatment (now known as KDAT)—that used heat to dry the wood out *before* it was shipped, with the result that more lumber could be moved per shipment.

A simple innovation on the face of it, KDAT turned the forest products industry on its head and introduced a new era in the production of chemically treated wood. In addition to cutting shipping costs, the KDAT process reduces the possibility of warping and produces a much better looking, higher-end piece of wood that today sells at a much higher price point than so-called "wet" lumber. Increasing the selling price of lumber was not the intended consequence; again, the motive had simply been to ship more wood more profitably. Still, KDAT allowed Cox Industries to position itself as a leader in the manufacture of high-end wood products. It's too bad

Bill Cox Sr. never attempted to patent it—KDAT is now widely used across the industry and is the standard for high-end wood. Had he done so, there is a high likelihood that Cox would have successfully differentiated and de-commoditized its business model at a much earlier time in its history.

Bill Cox Sr. wasn't necessarily thinking about what was best for the environment; he was a businessman and had to put food on the table for his family, so his foremost concern was with keeping the company profitable. Serendipitously, though, cutting costs here and there can also have an environmental benefit. KDAT is one example of this; the ability to ship more of a product per truck cuts down on fossil fuel use and greenhouse gas emissions. This type of serendipity—namely, an innovation introduced to increase profit margin having the unintended consequence of increasing the environmental sustainability of company practices—has appeared in other periods of Cox's history. So Cox has become more energy efficient and environmentally friendly as it has grown more profitable.

Fast-forward about fifty years: Cox, still family owned, implements a board of directors whose members are not in the family. Bringing in this outside expertise helps Cox become more self-conscious about both its innovation process and the environmental impact of its practices. I joined the board in this context, and I later joined the company as a specialist in innovation. This book is about the work that Cox has done in this area since that time. Given the promise and success of these efforts, Cox can serve as a model for other business leaders, especially those operating in smaller and mid-sized companies with limited resources, who are interested in implementing a proactive innovation program.

OBSTACLES TO INNOVATION

"Innovation" has without a doubt become one of the premier business buzzwords of the past decade, unseating the production-focused mantras of prior years like "lean manufacturing" and "total quality management." Companies are typically driven to innovate by one or more of a number of drivers. These include, of course, improving business profitability and meeting changing market conditions such as competitive threats or changes in consumption, but they can also include improving health and safety, aligning with overall corporate mission and strategy, or complying with government mandates.

When it comes to implementing innovations, however, things are more easily said than done. Longtime practitioners of innovation, of which there are many, know well that building a company that has a true portfolio of promising ideas to tap into is quite a challenging task. Such a portfolio needs to include initiatives that represent different levels of company risk and return, ranging from what Harvard Business School professor Clayton Christensen calls "incremental" innovations—tweaks to an existing product or service—to "disruptive" innovations—changes in strategy with the potential to transform markets or business models.

Cox had a history of innovation, but it was fortuitous—a kind of "ready-fire-aim" approach that had some key successes but could not be reliably depended upon to keep moving the company forward. The innovations that resulted were one-offs. It's nice to have that kind of luck; but as our recent history shows, it's better to have a well-defined process. The brainstorming process that led to a particular breakthrough in one area might produce other advances in the future or in other areas.

So, our approach to innovation became much more targeted, and we focused on developing a more formal innovation process for the purpose of developing a portfolio of ideas that are integrated with our business strategy and that are possible to implement systematically. The process we developed, which I describe more fully later on, is sequential, starting with identifying areas that are ripe for innovation, then carrying ideas through a strategic business analysis, and ending up with executable steps to take.

Such a process is especially necessary because disruptive change doesn't come easily at a company like ours, which faces challenges with regard to innovation simply due to our industry and our business profile. Cox is a mid-sized, family-owned, third generation manufacturing company, and it competes primarily in a consolidating commodity market—although stable, this market offers competitors only minimal growth opportunities year over year. Changes in market share within our industry typically arise from either acquisitions or the willingness of one competitor to underprice another to get more business.

Companies that fit our profile tend to be fairly risk averse; available resources to invest in innovation are usually limited, and are instead directed toward acquiring competitors or some other form of near-term financial win. In this environment, it is sometimes difficult to see how stretching the business beyond the status quo will be beneficial. However, this type of consolidation typically signals a mature industry attempting to protect a dying business model. This makes disruption all the more necessary—the sooner a company in this type of market faces up to the simple fact that innovation is a requirement for future organic growth, the better off it will be.

OBSTACLES TO INNOVATION:

1. **A company culture that resists change.**

   The workforce needs to be ready and willing to engage in and accept changes in the business.

2. **Leadership that does not support change.**

   Company leaders need to be willing to invest in and support innovation from a budgetary, resource, and timeline standpoint.

3. **Lack of resources to implement innovation wisely.**

   Companies need to strategically mobilize staff and resources in enacting an innovation process.

4. **Lack of awareness of customer needs.**

   Especially in commodity markets, companies must deeply understand the needs of their customers, who may not be looking for product-based innovation.

5. **Lack of strategic thinking about innovation.**

   Companies need to avoid implementing one-off or random innovations; instead, they should make sure their innovation program connects with their overall business strategy.

Still, the obstacles are considerable. Manufacturers that fit our profile often have an employee base with a fairly conservative mentality: "This is how we have always done things, and we have no intention to alter our processes or operations." This attitude is understandable: employees who have been doing one kind of work for many years have a type of ingrained expertise, or what some term "tribal knowledge," that really carries weight—especially if they are inclined to think about leaving the business if threatened by big changes. In the case of younger workers, workplaces can breed a culture of cautiousness or risk aversion, where surviving—keeping your job—by continuing to do things the way they have always been done trumps attempts to disrupt things or stand apart.

The challenge with family-owned companies often follows a similar pattern. Frequently these companies have a sort of "heritage" mentality that makes both leadership and employees push back against efforts to disrupt the status quo. Passive family owners, in particular, may not see the value in innovation. They are focused on the income the company is generating for them here and now, and are not interested in investing in some unknown future model.

Also, due to Cox's size, we exist in a kind of no-man's-land between, on the one hand, massive companies that can afford to invest millions in an R&D department staffed with scientists devoted entirely to exploring new possibilities, and on the other, nimble start-ups with the mentality that they have to be innovative in order to continue to exist. Large companies on the whole put less risk on the line when they devote resources to R&D, and they can operate against much longer timelines for execution, while start-ups are "all in" on innovation, making the long play of being acquired by a larger company. A mid-sized company like Cox operates with a different level of risk tolerance than both of these, so that gambling on the success of a long-term innovation project looks much less attractive.

Further, in a commodity market, the customer is typically not looking for innovative new products. They want the tried-and-true product, which is essentially the same from seller to seller, at the lowest price. "Cheaper" is better than "new and improved." For that reason, it can be difficult to get a return on investment on innovative new products or product features.

We found this out from some of the ethnographic research we did in our early explorations in transforming the utility pole, where we followed utility crews around on the job in search of ways to make the actual pole shape different based on how they would install it. At the end of the day, though, while the crews might like our ideas for making their jobs easier, the people who actually buy the poles really only care about price. Efforts to improve the poles themselves may not ever be compensated at all.

Put all of these factors together, and you get a fairly complex challenge to innovation efforts on the part of a company like Cox. As I mentioned, Cox had historically been a very innovative company, and the beauty of its more transactional, nonstrategic, "ready-fire-aim" approach was that it could move on an idea quickly; if the idea didn't work, it wasn't disastrous, because the company hadn't bet the farm on it. The downside of that approach is that instead of having innovation that fits into a broader business strategy, where we can determine exactly what resources and level of risk we're willing to put in to make the idea succeed, we have initiatives that are more like "ideas du jour," one-offs that are the product of someone seeing an opportunity but not really scouring the market to find out what needs to go into making that opportunity happen.

For instance, when I started at Cox, the company was using its lumber to make and sell outdoor furniture. The thinking behind this was basically "Why not? We have the wood, and it's easy enough to put together." Unfortunately, no one had really looked long and hard at what it takes for a company to succeed long term in the outdoor furniture market. In particular, it takes more than just putting pieces of wood together; you need a product design team and people who really understand what outdoor furniture buyers want—neither of which we had at the time, and neither of which seemed, upon reflection, to be worthwhile to invest in. In the absence of a broader strategy, ideas like these will never be more than incremental innovations at best, and dead ends at worst.

REQUIREMENTS FOR SUCCESS

A company that wants to be consistently innovative, then, needs a strategy for making innovation viable over the long term. For such a strategy to get off the ground and be successful in a company like Cox, certain factors need to be in place.

REQUIREMENTS FOR SUCCESS IN INNOVATION:

1. Alignment among all stakeholders.
2. Integration of innovation into overall business strategy.
3. A structured innovation process.
4. An internal culture of innovation that celebrates small wins.

The first of these is alignment among and support from all key stakeholders: the CEO, board of directors, leadership, staff, and family shareholders. These parties need to be aligned on the value and priority placed on the innovation function within the company, including expectations regarding factors like budget and time horizon.

Those of us pushing for innovation at Cox have been fortunate in this regard: the leadership has really gotten behind the effort, realizing that while there may not be an immediate impact on business performance, they have an opportunity here to shape the market rather than passively watching the market change around them. The Cox family, including the passive shareholders, have blessed us with a great amount of latitude in investing in innovation over the long haul. In fact, Cox is currently the only company in the wood treating industry with staff who are solely devoted to innovation.

This is not to say that there hasn't been skepticism from some of these parties. This is a hurdle for any company looking to enact change—some parties will inevitably see the effort as an unnecessary risk with an unlikely return on investment. This is the understandably cautious, conservative impulse of the mid-sized, family-owned company. This is especially true in the case of truly disruptive innovations; the impetus toward changing the business or moving into a whole new market can easily be met with an attitude that says, "We don't do that here." The key is to show the skeptics that the ability of the company to remain profitable while also growing and remaining dynamic in the market—controlling its future rather than having the future control it—is going to rest on how it utilizes the innovation process.

> ***This is a hurdle for any company looking to enact change—some parties will inevitably see the effort as an unnecessary risk with an unlikely return on investment.***

This consideration connects this first factor to the second requirement for success: integration of innovation into a broader, well-formed business strategy based on an examination of the likely impact of longer-term cultural, demographic, and market trends. Remaining competitive requires having some sense of where the market is headed and trying to get out ahead of those developments.

Cox reviews and updates its business strategy regularly, and it has become clear that considering the impact of these trends on business strategy is a pathway for developing potentially transformative innovations.

So, innovation needs to be part of a structured approach to achieving company strategy; it cannot be a standalone function that generates ideas in a vacuum, nor can it be focused on merely solving local problems. Convincing the skeptics involves awakening them to the idea that if we don't actively pursue innovation, then we are missing an opportunity to stay competitive on a longer-term basis. This is where trend forecasting comes in, which in Cox's case allowed us to identify an unmet need that was only becoming more urgent. I'll discuss this issue in much greater detail in later chapters, describing how venturing into a new model for utility pole disposal ultimately became the gateway for a new approach to pole production itself.

The beginning of Cox's innovation process involved developing what we called a "Four in Four" strategy, where we mapped out four major business initiatives or changes that we believed should take place over the course of the next four years to drive our business forward. We update this plan regularly, and it serves as a road map for where we need to be focusing our efforts. It is simply a way of fitting our innovation process within the broader strategic plan for the company.

In the commodity market we operate in, looking at things from the perspective of the customer—utilities in particular—made it clear that product-based innovations (i.e., making a new and better utility pole) were not going to add value for us; in cases like that, innovators have to turn to ancillary services based on their product, expanding their business model to include a service component. In our case, this meant creating new businesses under the Cox umbrella. While I discuss our pole disposal business, Cox Recovery, in depth in later chapters, the first development actually grew out of our exploration of the value of using radio-frequency identification (RFID) technology—a method for tracking items, similar to bar codes—to tag and track the poles we manufactured.

No one else in our industry was using RFID at the time. We started by implementing RFID in our own plants for internal inventory purposes, but we soon realized that this same technology would also allow utilities to better track, inspect, and maintain the poles once they were put up in their service areas. This led us to form a software company, Sustainable Management Systems (SMS), that essentially sells the capability for utilities to more quickly and accurately maintain their inventory of poles in use. Rather than reinventing the utility pole, SMS just attaches a service that provides added value for our utility customers by allowing them to move away from traditional paper-and-pencil inspection of poles. This is a great example of how we were able to get out ahead of the market by forecasting trends and needs.

The third requirement for success is actually having a true innovation *process*, so that we don't just revert to a transactional, opportunistic, one-off approach—yet not having a process so rigid or rigorous that it inhibits quick decision making and execution. This requires a Goldilocks-type approach—just the right amount of process to minimize risk, maximize strategic benefit, and maintain a degree of nimbleness. With the right process in place, the outcome will be something that has gone through some degree of vetting that warrants continued investment in it.

While I describe Cox's process below, it's important to emphasize that it is not a one-size-fits-all model. The literature on innovation provides countless examples of practitioners who have each developed their own specific innovation methodology. Each has merits and is worthy of consideration, but none are surefire successes in every case. The type of innovation effort that will work best for your company depends on your company's unique characteristics and circumstances, the challenges it faces, and a convergence of multiple factors including company culture, available resources, market conditions, and performance expectations.

> ***It's important to emphasize that it is not a one-size-fits-all model.***

Fourth and finally, an organization aiming to innovate needs to focus internally on small steps and small wins prior to engaging in any really disruptive or transformative changes. Generating examples of measurable success and advances on a small scale can visibly demonstrate and validate the importance and value of innovation, for leadership and employees alike. This both helps grow senior level support for the innovation process and sows the seeds of cultural change among the employees. One step that company leaders can take to foster innovation is to lead by example. Good leaders don't just "produce" innovations and push them onto employees from the top down; they enable innovation by pushing boundaries and asking provocative questions of their employees, looking for the latter to generate innovative ideas from the bottom up.

For example, to give our employees a sense of what we were up to with this whole push for innovation, Cox implemented an open innovation program called IdeaStream, which is basically an employee crowdsourcing process where employees are incentivized to offer up their own new ideas regarding any aspect of the company: products, services, operations, and so forth. On a regular basis, a committee evaluates all of the employee ideas and votes to implement any that are feasible and potentially valuable with a reward for the employees whose ideas get picked.

IdeaStream has provided the company with some valuable employee-based innovations, particularly in the areas of operations and final product quality. For instance, one of our employees who works on the front line handling poles at our facility in Vance, Alabama, drew up a detailed blueprint for a safety guard to control poles on the pole classing deck, which has not only increased safety but also reduced downtime. Another frontline worker, in Leland, North Carolina, designed and handcrafted a framing block to guide the direction of the drill bits used to drill framing holes in the poles, making it easier to produce consistent quality. We encounter these manufacturing areas—the pole classing deck and the framing yard—again in the next chapter, in the discussion of the life of a pole.

Perhaps most importantly, from a cultural point of view, IdeaStream has helped engage the organization on the importance of innovation to our future business success.

BREAKING DOWN THE PROCESS

As I mentioned, Cox's approach has been to implement a sequential innovation process that can reliably generate opportunities that can be acted on. The sequence begins with a meeting of what we have called a "forward-thinking group," consisting of employees who would be able to envision the future of the business over the

> ***Cox's approach has been to implement a sequential innovation process that can reliably generate opportunities that can be acted on.***

next five to ten years. With the expert help of a highly regarded innovation consulting firm, Vincent and Associates, the group first convened in December 2012 to really start framing the market assumptions that would affect our business going forward.

We presented the group with about thirty prompts regarding potential scenarios and problems over that time frame; the prompts were designed to make the conversation more guided and constrained than a free-form brainstorming session. We had also chosen to focus exclusively on the industrial and utility sales side of the business, as it offers the highest margin and thus is better able to generate enough profit to invest in innovation.

The prompts, focused within the context of our Four in Four strategy, addressed such issues as the future of our raw material supply. The growing wood pellet industry, for example, has introduced some market dynamics that did not exist ten years ago: owners of forest land can now profitably cut their trees down at a faster rate to serve this market, since smaller trees are more

PROCESS

1. A forward-thinking group meets with a facilitator for a prompt-based discussion to locate areas ripe for innovation.
2. The outcome of this meeting is an agreed-on set of four or five areas that present particularly promising opportunities.
3. These promising areas are evaluated against one another, a business case is built for each, and the cases are presented to the senior leadership of the company.
4. The company focuses solely on the single area that the business case was strongest for.
5. Unused initiatives become part of the company's innovation portfolio for possible use at another time.

in demand. This means the potential exists for a decreased supply of the very tall, thick trees required to make utility poles.

We also presented prompts concerning pole design and different modes of chemical treatment, and conducted a very lively debate about the future importance of sustainability, including the question of what to do with the wood waste once a pole has reached the end of its useful life span. The traditional option, a landfill, is becoming increasingly unavailable; it also provides no additional environmental value of any kind at the point of disposal, when in fact that value is there in the wood to be taken advantage of.

The forward-thinking group was tasked with exploring the trends and potential opportunities in our industry and in our market, to locate areas that were particularly ripe for innovation. The natural inclination at this stage is to try to develop solutions to perceived problems, but the aim of the forward-thinking group was simply to identify target areas we could train our attention on. In what areas do we see signals telling us there is a problem that needs to be solved? We should only start thinking about solutions after we figure out what problems are most promising.

The outcome of the forward-thinking group's meeting was an agreed-on set of four areas that presented particularly promising opportunities, ranging from coming up with a new, more eco-friendly treatment preservative to the disposal of wood waste. We then had to evaluate these against one another, build a business case for each, and present the cases to the senior leadership of the company. The goal of formulating the business cases was to do our due diligence regarding whether a particular avenue was worth pursuing from the point of view of the market, whether we could formulate a clear vision around that avenue that would motivate others in the company, and whether we had the resources and capability to execute in that area.

Because we are a company with limited internal resources, we recognized at the outset that we wanted to focus solely on the single area that the business case was strongest for rather than spreading ourselves too thin across multiple initiatives. The area of used pole disposal was the clear winner, for a number of reasons that I will discuss in the following chapters. The other initiatives developed by our forward thinkers were sidelined for the time being, though they remain a key part of our innovation portfolio, while we focused on trying to find opportunities to help utilities create a more sustainable model for their disposal practices.

This goal, the pursuit of which makes up the story of this book, was the impetus for the creation of Cox Recovery, a subsidiary of Cox that we established to help utilities find solutions to the question of where and how to dispose of old poles. The following chapters present a deep dive into this particular area of innovation, starting with the importance of tackling product innovation from the point of view of the product's whole life cycle, from cradle to grave, and then explaining the thinking behind the development of Cox Recovery. The story culminates with a discussion of the new frontier that Cox is exploring in working to develop the capacity to transform wood waste into a liquid fuel source that can then be used in the chemical treatment of new poles—turning the linear life cycle of the product into a closed-loop process. The promise of these developments exemplifies the value and validity of the type of innovation process and approach discussed in this chapter.

For smaller and mid-sized companies, corralling the resources to build a lasting innovation effort is not always an easy task, and stumbling blocks will undoubtedly pop up along the way. However, the potential upside for such a company, especially one operating in a commodity market or a consolidating industry, is that a structured innovation effort can truly become the impetus for transforming how the business operates, fueling sustainable business and disrupting commodity markets—and bringing value not only to customers, but to the general public as well.

CHAPTER 2

Life Cycle Thinking

When was the last time you really took a hard *look* at a wood utility pole? That's not something most people do very often, at least if they are considered sane, are of average IQ, and don't work in either electric utilities or pole manufacturing. The poles are simply a ubiquitous feature of our contemporary environment. It's estimated that there are over 150 million poles in use in the United States today—nearly one utility pole for every two people. The bulk of them (at least 60 percent) have been standing where they are for over thirty years, having been installed during the massive build-out of suburbia that occurred in the decades following World War II.

But most people, at least the ones I know, don't think about things like: *How long has that pole been there? How long will it stay there? Where did it come from, and where is it going?* Even our own customers, which range from the largest utilities in the country to small municipalities, may overlook the important role that poles play in our society. These rather mundane products truly are the backbone of our electrical system, without which many of the conveniences and technologies we enjoy today would not be available to us. So, though the common wood utility pole may not be a topic of interest for many and may be forgotten or simply taken for granted, the job of the innovator is to think about the questions that others don't ask, and hopefully develop solutions that answer them.

> ***The job of the innovator is to think about the questions that others don't ask, and hopefully develop solutions that answer them.***

As a forest products company, Cox harvests trees, produces and treats the poles that are made from them, and sells those poles to utilities. Thinking about any product—in this case, the utility pole—across its entire life cycle, from cradle to grave, opens up new opportunities for innovation at every point. Most industries refer to this as product life cycle thinking. In our instance, the "cradle" is the planting of the tree that ultimately grows into the pole. The "grave," of course, is how the pole is disposed of once it has reached the end of its useful life span. It is here that we found the greatest potential for our innovation efforts, especially when it comes to disruptive innovations.

In Cox's world, an incremental innovation might be something that impacts the performance of our existing pole product line—for instance, a new way to limit the growth of vegetation around poles or a pole design that better accommodates the growing need for mobile broadband and Wi-Fi connectivity in rural America. Both of these examples, incidentally, were real attempts to meet market needs that emerged from our innovation process. On the West Coast, unchecked growth of vegetation increases wildfire risk. At the same time, the labor to install a single pole costs utilities around $1,500, so a pole that's less susceptible to damage from wildfires could lead to a huge cost savings for utilities.

These incremental innovations are important company assets to incorporate into our portfolio, but for our innovation work to be truly unique, meaningful, and disruptive, it must have a simultaneous positive impact on multiple stakeholders. We can't limit our thinking to simply what's best for Cox, or even for our customers. Instead we must address broader issues that affect *their* customers, our communities, and ultimately our planet. For us, focusing our innovation work on the used pole graveyard provides a platform to become disruptive, particularly if we're able to find an opportunity to create greater value—environmental value, for sure, but environmental value that is, at the same time, bottom-line value, and which is beneficial to a number of stakeholders.

We can't limit our thinking to simply what's best for Cox, or even for our customers. Instead we must address broader issues that affect* their *customers, our communities, and ultimately our planet.

THE LIFE OF A POLE

Before focusing solely on how to innovate in the area of pole disposal, it's important to better dissect the true life cycle of a pole, from initial planting of the tree, through manufacturing, and then ultimately to disposal. In this way, we can determine whether an innovation that is designed to have an impact on one part of the life cycle might also have a positive effect on others. As we'll see later, while much our innovation work focuses on end-of-life solutions for poles, we have also found this same effort positively touches other parts of the pole life cycle.

We can determine whether an innovation that is designed to have an impact on one part of the life cycle might also have a positive effect on others.

So, what shape does our product life cycle take? In the case of Cox, we work predominantly with southern yellow pine trees, and less frequently with Douglas fir trees. Generally speaking, these are the two most common tree species used for poles in this country, depending on customer preferences. With manufacturing plants located throughout the southern United States, Cox is fortunate to reside in what might be termed the "Saudi Arabia of southern yellow pine." This species is in plentiful supply and is likely to remain so with proper stewardship of the forests.

We have a group of foresters on staff who work directly with private landowners who are growing and harvesting southern yellow pine. We work collaboratively with the landowners in the initial purchase, planting, and growth of the trees, as well as in the effective maintenance and management of their land. Our growers all operate in accordance with the standards imposed by the Sustainable Forestry Initiative, a nonprofit entity created to ensure that forests are managed and harvested in a sustainable manner that promotes longevity, productivity, and the overall health of forest resources.

Most of the plots of land where we harvest trees have multiple "cutting lines," meaning trees that are cut down and replanted at a certain age, both to create ongoing income for the grower and to reduce the potential effect of disease impacting future cuts. For instance, those who are harvesting and selling the wood may make a cut of a subset of the trees after they have grown for ten years. This offers wood for use in industries that can work with smaller trees, like the paper and pellet industries.

For utility poles, we typically take the last cut, consisting of the oldest trees. After a line of trees has grown for about thirty years, we cut it down and shape it into poles of varying lengths. We harvest based on a profile of what we need in terms of the tree size (both height and diameter). Within the industry, this sizing is referred to as the pole "class." There are thirteen main classes of poles in use today, with multiple variations within each class. Different customers (in this case, different utilities) have different needs for poles based on the situations where they're being used and what sort of equipment they need to be able to bear the weight of.

After the trees we have selected are cut, we help the landowner plant new trees in such a way that we end up with a roughly two-to-one ratio of new to old trees—essentially, we are engaged in continuously reforesting the land; this in itself is a microcosm of thinking from the point of view of the product life cycle, which helps us maintain and grow the stock of trees that provide the raw material for our products.

The next step in our product life cycle is the actual manufacturing process that turns a tree into a usable utility pole. The trees we have cut are taken to a peeling plant, where the bark is stripped off and the diameter of the tree is further shaped and smoothed prior to being treated.

Next, the poles are taken to a treatment facility, where they are placed on a rail tram-like device and rolled into a giant tube called a treatment cylinder. The cylinder is filled with a chosen chemical preservative, and a liquid is added that helps to serve as a "carrier" for the preservative as it impregnates the wood. The most common carriers are water and diesel fuel; which is used depends on the type of preservative being applied. Through a vacuum pressure treatment process, the poles soak up the liquid, and the chemical preservative with it. The poles are then removed from the tube, the chemical is pumped into a storage facility for safe handling, and the poles are then either air or kiln dried. Lastly, the poles are moved to what's termed a "framing yard," where workers drill holes into the poles at specified locations to allow the utility to fit cross arms onto the pole. These cross arms serve an important role, allowing for power lines to be successfully strung from pole to pole.

The final output of this process is a pole that can be sold to a utility and put to use in the field. Just how long the pole will last in the field depends on multiple factors, including which preservative it has been treated with (each chemical has its own unique characteristics) and the climate it will be subjected to. In general, poles placed in a drier climate that's free from temperature extremes last longer than those in more humid climates or climates that experience wider temperature fluctuations and are more likely impacted by insect-related damage. Perhaps the biggest impact on pole life span doesn't come from the local climate, but rather from the commercial development that surrounds it. In more rapidly developing areas, utilities may have to regularly find new locations for their poles due to road expansion or changes; this can considerably shorten a pole's life span, as liability issues most often preclude a utility from reusing a pole once it has been removed from one location. Generally speaking, with the proper matching of chemical and climate, a pole's useful life should span up to thirty or forty years if it is left relatively undisturbed.

WHY CHEMICALS?

One of the most frequent questions I hear from people outside of our industry, particularly those with environmental concerns, is "Why do you have to put chemicals on the poles to begin with?" The common belief is that these chemicals, which are really derivatives of pesticides, will have significant negative effects on both humans and the environment. I grew up in Oregon, one of our country's most environmentally conscious states, and from the earliest years I was made very aware of how our actions can have either a positive or a detrimental effect on the environment. Consequently, I am well versed in how and why chemical usage is generally viewed as a big negative.

Perhaps it is important first to understand just how much chemical is used in making a pole. All major pole producers abide by manufacturing standards developed by the American Wood Products Association (AWPA). This association, composed of scientists, academics, and industry personnel, collectively establishes the amount of chemical required during the manufacturing process to maintain the proper functioning of the pole. Per AWPA standards, the typical pole is impregnated with chemicals that, depending on tree species, penetrate less than four inches of the entire diameter of the pole. Given an average-sized pole based on industry standards, this means that less than 5 percent of a pole's entire mass receives chemical treatment, a relatively small fraction.

Still, that doesn't answer the question of "why?" While there are a handful of different types of chemicals used to treat wooden utility poles, they all have the same effect: they extend the pole's life span by protecting it from things like fungi or termites—basically, any organic thing that might feed off the wood and lead the pole to rot out and fall apart.

If the poles went untreated, they would rot out so quickly that they would become much more likely to break, especially during inclement weather, and it would become extremely difficult for wood vendors like Cox to keep up with the demand for new poles on the part of utilities. In storm-prone areas like the South, poles would most likely have to be replaced after only five or ten years of use. Imagine, then, the amount of timber that would need to be harvested to deal with this issue—roughly four times the amount that is harvested today. Add to that the changes in environmental impact caused by the decline of sequestered carbon in the forest, as well as the effects of logging and manufacturing to increase output, and the overall result is that chemical usage actually does more to help preserve our environment than to harm it.

On a more practical level, the cost of maintaining a utility infrastructure with untreated poles is much higher than utilities can sustain—they not only have to purchase the poles, but they also have to pay for the labor to put the poles up and string the lines. Costs would go through the roof if they had to do this every few years, and most likely ratepayers like you and I would bear the burden through increased utility charges.

Still, while chemical treatment can be shown to be beneficial, that hasn't dissuaded us from considering another area that we believe is ripe for innovation: a cleaner, more eco-friendly preservative that could replace the currently used chemical preservatives while still allowing poles to maintain a long service life. We work with our chemical suppliers on this, and it is our hope that such a breakthrough can and will occur, but for now our innovation process and our sustainability initiatives focus on the issues that are more directly under our control; a major one of these is carbon footprint, an area that our industry is powerfully positioned to affect.

LIGHTER FOOTPRINT

An unanswered question from the previous discussion of chemically treated wood has to do with the potential value of using non-wood utility poles. There certainly are utility poles made of other materials—concrete, galvanized steel, and fiber-reinforced composite being the most common. Would these not be better options, particularly with regard to their associated environmental impact?

From the point of view of sustainability and carbon footprint, the answer for each of these alternatives is "no," for reasons that, again, have to do with the product life cycle. From the life cycle perspective, wood utility poles have an enormous advantage over other material types in the area of environmental sustainability and carbon footprint. To dissect this issue further, the Treated Wood Council, in partnership with engineering firm AquAeTer Inc., conducted a life cycle assessment of utility poles made from various materials, taking account of the environmental impact of the production, use, and disposal of these pole types. The assessment found that treated wood compares favorably against galvanized steel, concrete, and fiber-reinforced composite poles along several environmental metrics, including

> ***From the life cycle perspective, wood utility poles have an enormous advantage over other material types in the area of environmental sustainability and carbon footprint.***

greenhouse gas emissions, fossil fuel use, ecological toxicity, and water use.[2]

Much of this advantage is due to the fact that with forest products, we are dealing with an organic raw material. Contrast the wooden pole with galvanized steel from a life cycle perspective, for instance: To provide the raw material for the latter, steel has to be produced from iron ore and treated with zinc (i.e., galvanized)—both carbon-costing processes that draw on nonrenewable resources. With a wooden utility pole, part of the production of the raw material is the thirty years of wood growth in the form of a tree, starting with its planting. The raw material is drawn from the soil, while the production process runs entirely on solar power, so to speak. This growth process actually *removes* carbon (in the form of carbon dioxide) from the atmosphere, rather than contributing to carbon emissions—a remarkable side effect of using this organic raw material. The so-called "carbon sequestration" process generates a "carbon credit" early in the life cycle that allows the final product to have a much lower carbon footprint than competing products.

During the forty or so years that the pole stands, another tree will have grown in its place—canceling out the carbon release from the pole at end of life; the whole process thus potentially lowers carbon

---

2 Several studies have shown this including:
Christopher A. Bolin and Stephen T. Smith, "Life cycle assessment of pentachlorophenol-treated wooden utility poles with comparisons to steel and concrete utility poles," *Renewable and Sustainable Energy Reviews* 15, no. 5 (June 2011): 2475-2486, https://doi.org/10.1016/j.rser.2011.01.019.
Richard Bergman et al., "The Carbon Impacts of Wood Products," *Forest Products Journal* 64, no. 7/8, https://www.fpl.fs.fed.us/documnts/pdf2014/fpl_2014_bergman007.pdf.
"Conclusions and Summary Report: Environmental Life Cycle Assessment of Ammoniacal Copper Zinc Arsenate-Treated Utility Poles with Comparisons to Concrete, Galvanized Steel, and Fiber-Reinforced Composite Utility Poles," AqaAeTer, Inc., last modified 2012, http://www.conradfp.com/pdf/acza-poles-ca-summary.pdf.

dioxide emissions rather than contributes to them. The carbon emissions that do happen in the process of producing a wood utility pole occur at the level of preparing the pole for use, treating it, and transporting it. Apart from the chemical treatment, these operational tasks will likewise occur with a utility pole made from any competing material.

This brings us to the last stage of the product life cycle: disposal, which is the focus of the innovation efforts discussed in this book. Whether due to a pole reaching the end of the roughly forty years that it can function or due to its being removed to make way for something like a widening road, the utility next has to get rid of the pole in some way. Typically, it is moved to a landfill and left to decompose. Our research indicates that over half of all disposed poles in the United States are landfilled. At this stage, the carbon stored in the wood is released into the atmosphere. That is one way to transform a utility pole. This stored carbon can, however, be used to generate energy after the end of the pole's life as a utility pole—this is the basis for other ways to transform a pole, the alternative types of disposal discussed later.

The decomposition of the pole at the end of its life, and the resulting emission of carbon into the atmosphere, is wasteful. For that reason, the way we saw it, the traditional model of shipping old poles off to a landfill was offering no added value, from an environmental standpoint, for any of the stakeholders involved—whether the utilities that are our customers,

> ***The way we saw it, the traditional model of shipping old poles off to a landfill was offering no added value, from an environmental standpoint, for any of the stakeholders involved***

consumers in general, or environmental groups. We were motivated to look into this because our customers—utilities—are beginning to be engaged in and focused on issues of sustainability. Some are looking at more aggressive approaches simply because government regulations are starting to take effect. Others are being driven by their own internal goals to manage their waste stream in a more environmentally friendly manner.

The question posed to our innovation group, then, is how to generate environmental value downstream in the process, at the time of disposal, when the traditional model currently generates zero value for anyone involved (except the landfill owner). According to some, this will only happen when the government intervenes by force of regulation. My hope, and the hope that motivates this book, is that we will not have to let it get to that point; we can offer up a private industry, value-adding solution that works to everyone's mutual benefit.

CHAPTER 3

Landfill: An Uncertain Future

All productive human activity produces waste in some form, and—while it may not seem like a profound philosophical question—we have always had to ask ourselves: "What do we do with all of this stuff once we're done with it?" Waste is often material we don't really want to keep around, but we can't help producing it, and it has to go *somewhere.*

The tendency to just dump all of that waste material in one designated spot gave rise to the evolution of the open city dump, which was the predominant solution to this conundrum up into the twentieth century. Dumps were largely unregulated; they bred pests and, since much waste is toxic, allowed poisonous materials to leach into the ground and affect the water supply and the broader environment. The landfill developed in the early twentieth century as a cleaner, safer alternative and would eventually grow to replace the city dump. The idea behind the landfill is essentially to isolate the garbage (known as "municipal solid waste") in a confined space, control leaching and gas emissions, and cover the surface with soil on a daily basis. In the United States, landfills are subject to stringent EPA regulations.

The first modern landfill started operating in 1937 in Fresno, California. In the years since, most Americans have come to take the facilities for granted, simply throwing waste away without thinking about where it is going. These facilities, however, are running out of space; across large portions of the Northeast and especially the upper Midwest, there is less than twenty years' worth of useful life in the landfill space available. Many of the country's largest landfills have already closed their doors. Staten Island's Fresh Kills, for instance, has been closed for fifteen years and is being made into a public park (more than twice the size of Central Park). Space started becoming problematic there at least as far back as 1993, as discussed at that time in an article in *The New York Times*.[3] Puente Hills, outside of Los Angeles, was the largest operating landfill in the United States until it closed in 2013.[4]

Utilities have long operated on the assumption that there is essentially limitless landfill space.

Like almost everyone who produces waste and has to get rid of it, utilities have long operated on the assumption that there is essentially limitless landfill space. In the eighty years since the first

3 Matthew L. Wald, "Buying Time by Buying Space on Mount Garbage; Engineers at the Fresh Kills Landfill Work to Squeeze in Every Last Inch of City Trash," *The New York Times*, last modified 1993, https://www.nytimes.com/1993/05/11/nyregion/buying-time-buying-space-mount-garbage-engineers-fresh-kills-landfill-work.html.

4 This article is a fascinating tour, with photos, of Puente Hills, with some good background information on how well-run landfills operate: Geoff Manaugh and Nicola Twilley, "Touring the Largest Active Landfill in America," *The Atlantic*, last modified April 5, 2013, https://www.theatlantic.com/technology/archive/2013/04/touring-the-largest-active-landfill-in-america/274731/.

landfill was created, the vast majority of utilities have not changed their waste disposal operations, with most relying heavily on landfills to take in their used poles. The traditional model for most utilities is to cut up and pile the used poles in a dumpster. Once the dumpster is full, they call a local waste management company to come pick it up and empty it into a landfill. Every year, about four million tons of treated wood utility poles are disposed of, and at least 60 percent (or 2.4 million tons, a conservative estimate) is landfilled.[5] The space for these poles, though, is limited; this solution is unsustainable.

REMAINING US LANDFILL LIFE

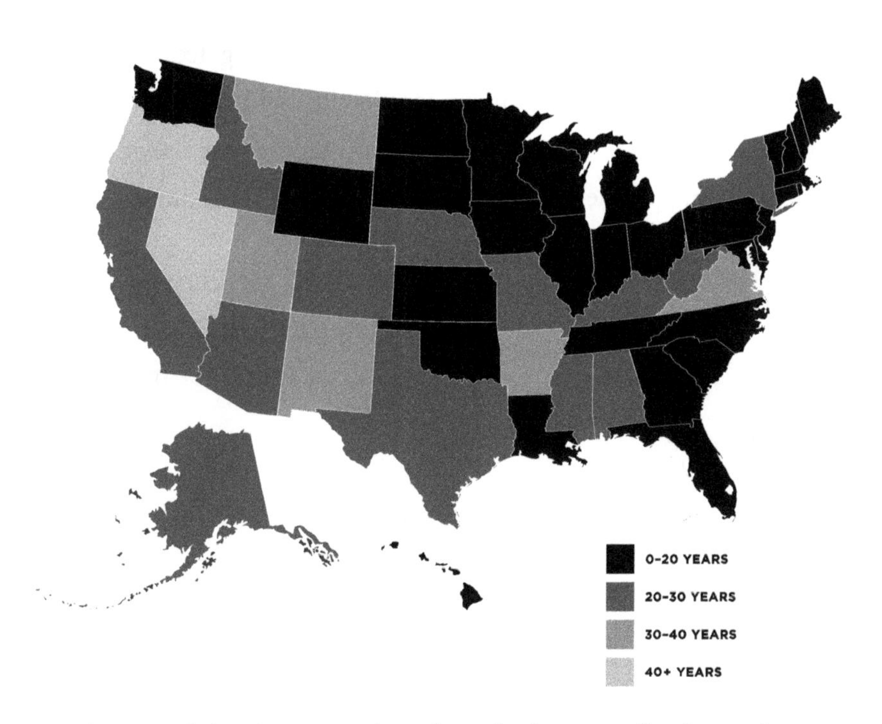

Projected developments based on the history of pole production will only exacerbate the problem. Historically, on average about 3

5 "How Utilities Dispose of Poles," Cox Market Research, last modified October 2013.

percent of the country's utility poles have been replaced annually. However, as I mentioned in the previous chapter, this history has largely been in the environment of postwar expansion of suburbs and the infrastructure development that went along with it. Much of that infrastructure is now reaching its age limit, such that the percentage of utility poles replaced on a year-to-year basis is going to start increasing. This, of course, means that the actual amount of pole waste itself is going to grow significantly in the coming years.

At the same time that landfill space is becoming increasingly scarce, it is also (and as a result) becoming increasingly costly. Of course, when space goes down, the price of entry goes up. Landfill operators charge what is known as a "tipping fee" for use of the landfill space, and as that space is becoming rarer, tipping fees are going up. As you may imagine, tipping fees in the Northeast and Pacific regions of the US are considerably higher than they are elsewhere—a particularly big problem for utilities operating in these areas.

AVERAGE MSW LANDFILL TIP FEES, BY REGION

REGION	AVERAGE TIPPING FEE		
	2016	2017	DIFFERENCE
Northeast (CT, DE, ME, MD, MA, NH, NJ, NY, PA, RI, VT, VA, WV)	$58.20	$67.27	+$9.07
Pacific (AK, AZ, CA, HI, ID, OR, WA)	$61.20	$60.20	-$1.00
Midwest (IL, IN, IA, KS, MI, MN, MO, NE, OH, WI)	$39.64	$45.84	+$10.63
Mountains/Plains (CO, MT, ND, SD, UT, WY)	$43.38	$45.84	+$2.46
Southeast (AL, FL, GA, KY, MS, NC, SC, TN)	$44.46	$41.01	-$3.45
South Central (AR, LA, NM, OK, TX)	$36.34	$36.94	+0.60
National Average	**$48.27**	**$51.82**	**+$3.55**

These factors add up to landfill being an increasingly unsustainable disposal option for utilities. In this chapter, I dig deeper into some of the problems that utilities are starting to face when it comes to the traditional model of pole disposal.

ARSENIC AND OLD WOOD

In addition to aging infrastructure, the production trends for poles made using different types of chemical treatments further exacerbate the landfill problem. The two dominant chemicals used today to treat utility poles in the United States are pentachlorophenol ("penta," for short) and chromated copper arsenate (CCA). A third, creosote, is found mainly in railroad ties and in utility poles in Texas, as the use of creosote largely grew out of the oil industry. Because of their chemical composition, both creosote and penta can legally be incinerated in many states, but CCA cannot, due to the presence of heavy metals. For that reason, penta- and creosote-treated poles are candidates for some of the alternative methods of disposal that I will discuss in Chapter 6; on the other hand, at this point, landfill is essentially the *only* disposal option for CCA-treated poles.

Over the last thirty or so years, CCA has been used more and more widely, and is now the dominant chemical used to treat poles. This is almost entirely due to price—CCA is the most economical choice, for a couple of reasons. First, it is important to note that the chemicals differ in how they are pushed into the wood—all of them soak in via some kind of liquid carrier, but the carrier for penta and creosote is petroleum based, while for CCA the carrier is water. Water is of course cheaper than the diesel required for penta (CCA's primary competitor). Second, the market for CCA is more competitive than it is for penta. Together, these factors lead to a cheaper product, which generally makes it the more attractive alternative to the buyer at the utility.

As a result, CCA has gone from being used in about 10 percent of poles manufactured twenty or so years ago to being used in about 50 percent today. So, while most of the poles that may be coming out of the ground *today* are creosote or penta treated, having been manufactured about forty or so years ago on average, the proportion of waste poles that are CCA treated is going to grow rapidly in coming years. Fast-forward to twenty or thirty years from now, and, without the development of any new disposal solutions, we're going to have a majority of poles coming out of the ground that can't go anywhere but the landfill—and there likely won't be any landfill for them to go to.

We're going to have a majority of poles coming out of the ground that can't go anywhere but the landfill—and there likely won't be any landfill for them to go to.

In many instances, the decision about where or how a pole can be disposed of is not in the hands of the utility, but rather lies with the regulatory bodies at the state and federal levels responsible for managing treated wood waste. So, in the early stages of our innovation process, when we were starting to do market research surrounding pole disposal possibilities, it became clear that one of our first conversations needed to be with the government bodies that control pole disposal, simply to understand their perspectives on this issue.

When it comes to the disposal of used wood utility poles, one of the most important governing bodies in our country is the Environmental Protection Agency (EPA). The EPA establishes regulations governing both the classification of waste (hazardous or nonhazardous in nature) and the applicable type of disposal option(s) that can be used. Decades ago, the EPA classified used treated utility poles as nonhazardous waste, thus allowing the material to be placed in landfills. In many instances, depending on state law, these must be lined landfills to prevent potential leaching of chemicals into the soil.

My first informal introduction to the EPA occurred a few years back at a Washington, D.C.-based conference on biofuels. The EPA typically sends a contingent of representatives to these conferences to act as the spokespeople for the government, attending meetings with industry leaders and sometimes making their own presentations. During the Obama administration, these groups were pretty active at these conferences, as there was fairly strong government support for building a renewable energy platform, including biodiesel and other alternative fuel sources. I was at the conference because Cox had just become aware of the potential for those operating in the renewable fuel industry to use treated wood waste as a feedstock for producing their products.

At the conference, I spoke with a couple of EPA representatives regarding Cox's interest in finding new methods for disposing of old treated wood utility poles. I must admit that I got somewhat puzzled looks in response; they were unclear on the connection between this interest and the biofuels conference we were currently attending. I realized soon that the issue was not on their radar at all, so I explained that, currently, the only place a lot of these poles could go was the landfill, and the landfills are filling up in a lot of areas in the country. They were certainly aware of the problem of limited landfill space, but were somewhat stunned by how restrictive the options are for how to dispose of these poles.

As I laid it out for the EPA, when you combine the challenge of decreasing landfill space with the growth of CCA-treated poles in this country, you begin to see how a perfect storm of sorts is about to unfold in the years ahead. We are manufacturing a pole that today has really only one disposal option: landfilling—while in fact we have less and less landfill space to work with.

On a broader level, I learned an important lesson in my meeting with the EPA. For corporate innovators working in industries that are in some way regulated by the government (and who isn't?), it's important to develop a shared understanding of the problems that need to be solved before you embark on potential solutions. In doing so, there is a far greater chance that you'll create an ally in your innovation process rather than a potential future stumbling block. Often this will take some extra time and energy to educate folks on issues that you live with daily while they do not. However, the longer-term upside is well worth the effort.

A CONFUSING MATRIX OF STATE LAWS

While the EPA governs the classification of pole waste as nonhazardous, regulations governing what specific disposal options are available locally vary from state to state. The result is that disposal solutions that are available in Massachusetts may not be available in Missouri. This is posing an increasing challenge for larger investor-owned utilities that operate on a regional or multistate level.

Compounding the issue of what can or cannot be used as a disposal solution are the seeming regulatory contradictions that exist within a given state. A rather extreme example of this situation involves California's public utilities. The disposal of utility poles (and other treated wood) is governed in California under a state law known as Assembly Bill 1353. This law, initiated in 2005, mandates that all treated wood, including utility poles, can only be disposed of in one of the state's forty-plus approved landfills. Failure to abide by this law can result in the state classifying the waste as hazardous, resulting in both significant financial penalties and environmental backlash for the utility. In effect, this law forces utilities to use landfills for pole disposal even though other locally available and more environmentally friendly options, such as the waste-to-energy facilities discussed in Chapter 6, are available for potential use.

> ***This law forces utilities to use landfills for pole disposal even though other locally available and more environmentally friendly options, such as the waste-to-energy facilities discussed in Chapter 6, are available for potential use.***

To further compound the challenges for California utilities in trying to abide with this regulation, certain municipalities, like San Francisco, have mandated that businesses achieve a zero-waste goal by 2020—meaning having no material end up in a landfill.

As part of Cox's innovation push, I met with utility executives in California to discuss possible disposal solutions, and I was stunned by the impossible situation that local mandates had put them in. These utilities would prefer to pursue an environmentally sound strategy to dispose of the poles, and have now essentially been ordered by their local municipalities to do so as well—but are at the same time hobbled by conflicting state regulations. In the meantime, as is true for a lot of utilities, those operating in California keep sending their old poles to the landfill. It's still the path of least resistance, even if it won't be in a few years.

To some extent, this problem can only be solved by changes in California laws, but it provides a particularly stark example of the paradoxical situation in which utilities can find themselves if they, or the lawmakers in the states they operate in, cannot find alternative disposal methods. Whether it is the result of a government mandate or they just get turned away from or priced out of landfills due to lack of space, utilities are going to have a major operational problem on their hands.

Misalignment between different states, as well as continual changes and uncertainty at the federal level due to changes in administration, make governments an unreliable partner in this area, so businesses have to take the lead in pushing for sustainable solutions that are economically feasible. Our aim must be to get ahead of the coming developments from the side of private industry, by first educating utilities on what disposal solutions are available to them so that a new mentality is created, one that will allow them to be flexible and competitive in the face of these developments. Fortunately, some solutions do already exist, and Cox Recovery was put in place to help utilities connect with them.

We are left, then, with dwindling landfill space, a patchwork of unpredictable state and federal regulations, a changing business climate where sustainability is becoming more of a focus, and the likelihood of a rapidly increasing waste stream with nowhere to go—seemingly a perfect juncture for us to begin further targeting our innovation efforts.

CHAPTER 4

A Winning Proposal

Looking back, it seems like a no-brainer. Even before we had put together the forward-thinking group as part of the innovation process I described in Chapter 1, we were catching wind of some increasing anxiety from our customer base. Landfill costs were going up, while at the same time state legislatures were handing down new rules about how to handle waste. The utilities were clearly feeling the pressure to find a more sustainable model for disposing of poles, but the options just weren't there; it was clearly an area that called for innovation, and no one else was really working on it.

So, how could we help? What needs could we immediately meet? These questions drove our innovation efforts and ultimately led us to establish Cox Recovery, a subsidiary of Cox Industries specializing in the recovery and disposal of old poles, reels, pallets, and pretty much any other wood-related waste. The development of Cox Recovery exemplifies the possible outcomes of the type of focused innovation program we had set out to implement. Arriving at Cox Recovery through the innovation process allowed us to overcome the challenges to innovation that an old-line, mid-sized, family-owned commodity manufacturer can face, and to meet all of the requirements for success I discussed in the first chapter.

While it seems like an obvious step now, we only committed to the Cox Recovery model because it came out the other end of the test of our innovation process with more promise and potential than any of the other ideas. This was the biggest initial outcome of our targeted and strategic innovation process—though, as I'll explain, it is only the beginning of an even more disruptive innovation path. From the initial impetus—responding to market needs and integrating with our Four in Four strategy—to the overall aims of the company, Cox Recovery illustrates the power of our targeted innovation process.

THE MARKET SPEAKS

Compared to the other areas of innovation that were the outcome of the forward-thinking group's meeting, moving into the pole disposal area wouldn't require an enormous change in how we conducted business, at least initially. This is a benefit for an old-line manufacturer, where it might seem that the way we have always done things was working just fine, so why try something new? Cox Recovery, though, required relatively minimal financial, logistical, and operational support to launch, so employees and leadership were less likely to resist.

The other options would have required much bigger investments and much more risk as far as what the market was willing to accept. For example, the idea of developing a more environmentally friendly form of chemical treatment, while beneficial, had less obvious immediate applications for us and seemed to have a much higher chance of just not going anywhere—not to mention the fact that we are not, at the moment, a biochem company, but rather a forest products company.

In contrast, logistically speaking, disposal was a natural area for us to move into, because we knew we could take control of it and have an impact, whereas the other areas of innovation further upstream in the product life cycle were further outside our domain of capability. Rather than being a drastic lateral move into a totally new area, the establishment of Cox Recovery cohered with our basic business model as it already existed, and with the services we were already offering. For instance, we had the trucking assets already in place that could pick up used poles and dispose of them. The thinking was basically this: "Well, we already make the poles, transport them, and drop them off; why don't we just pick up any used poles while we're already in the field and figure out the best thing to do with them, so that the utility doesn't have to?"

We were also confident of a return on our investment. A lot of innovations tend to be ahead of the market, and sometimes it will be a while before the market will accept them; but we could see that the market was likely to catch up with us quickly if we pursued this avenue. The early stages of the innovation process described in Chapter 1 involved our going out and doing market research among customers and prospective customers, to see what their pain points were and how we might be able to help address them. It soon became clear that there was a huge opportunity present in trying to find ways to help these customers create more sustainable business models for themselves. We kept hearing from our biggest customers—investor-owned, publicly traded utilities—about the greatly increasing scrutiny they were facing with regard to their environmental sustainability practices, so we knew that making these practices more sustainable from an economic standpoint as well would be very attractive to our customers.

We kept hearing from our biggest customers—investor-owned, publicly traded utilities—about the greatly increasing scrutiny they were facing with regard to their environmental sustainability practices.

This is in contrast to our RFID-based business, Sustainable Management Systems (SMS), which has now been in operation for several years. SMS was our first foray into the realm of a concerted innovation effort; however, this was before we had worked out the innovation protocol and process outlined in the development of Cox Recovery. That meant a lack of adequate preparation. As a result, when we first launched SMS, we failed to understand the specifics behind our go-to-market strategy. We erroneously thought the utilities were ready to adopt RFID into their business. They weren't. Manual processes and bar coding were working just fine in their minds. We also thought our SMS customer would be the same pole buyer that our sales force had been dealing with for years, thus giving us some added leverage from existing relationships. However, we quickly learned that the audience was not the pole buyer, but rather corporate IT and, in some cases, field operations. We also learned that a salesperson trained in selling wood is not likely to have the skills to sell software to an IT professional. The result of these shortcomings was that SMS stumbled initially and only gained traction once the market started to embrace the technology and once we teamed with other entities that were far more familiar, involved, and competent in the RFID market.

The Cox Recovery model, on the other hand, is directly aligned with a customer need. Presenting utilities with options for the disposal of their used wood poles solves a problem for them, removing a task from their operational workload and enabling them to more effectively and conveniently comply with regulatory requirements. We take the matter out of their hands.

THE RECOVERY MISSION

Utilities want to reduce waste; the EPA wants to reduce landfill usage; and everyone wants to find an environmentally friendly way of disposing of this treated wood waste. Our attempt to align ourselves with these interests meant that we did not simply aim to serve as a waste management service. Our message was not to say, "Don't worry about it, we'll load that stuff into a dumpster and drive it to the landfill for you." Our first focus was on finding things to do with the material *besides* just putting it in a landfill. As I discussed in Chapter 2, we are interested in creating *environmental value.*

> ***Our first focus was on finding things to do with the material besides just putting it in a landfill.***

We also aim to provide the utilities that work with us environmental reporting for sustainability purposes. This is another way we help fuel sustainable business. We help the utilities build metrics around their practices regarding how they dispose of waste, in particular helping them track their efforts to divert material away from the landfill and toward other, more environmentally friendly, methods (I discuss some of these in Chapter 6). So if they chose, for instance, a waste-to-energy solution, we would give them metrics on how much of their material is now generating X amount of electricity in a given area as a result of having diverted it out of landfill.

This can be a major benefit to a company like a utility. In most cases, a utility's sustainability efforts center predominantly around their energy production—i.e., to what degree are they relying on solar, wind, or other renewable resources to generate power. In fact, some utilities don't really do any kind of sustainability evaluation beyond the programs provided by the solar panel company they're contracted with. The raw materials they use to actually transport that energy might well not even be on a sustainability director's radar. Still, helping utilities report on their diversion of waste from landfill toward more eco-friendly alternatives also fuels sustainable business and enables a sustainability director to report good news for the company, which has a ripple effect through the whole organization.

Another key component of our mission is that we aim to offer what is essentially a turnkey service by leveraging the logistics network we already had in place for pole delivery toward implementing pickup and removal services. We started by telling our existing customers, "All you have to do is get the old long poles to the storeroom or other designated area—we'll come pick them up, and we'll handle it from there. You only have to work with us, not multiple vendors or contractors." We wanted anyone who contracted for our services to know that they would not have the hassle of a lot of the typical issues that go along with waste disposal, whether that is cutting up material to fit into a dumpster, contracting with a dumpster company to transport that material to a landfill on a regular basis, or paying the tipping fee to the landfill operator. We would streamline the process and make it very efficient for the utility from an operational standpoint.

While I discuss the pricing benefits of this type of disposal service in the next chapter, it is worth pointing out here that this type of turnkey solution creates a much more efficient cost structure for utilities. Pole disposal is the source of a lot of hidden costs in a utility's operations. First, of course, there are labor costs for keeping track of the dumpster delivery, tracking how much wood goes into the dumpster, and actually cutting up that wood and putting it in there. With a lot of material coming in and going out of a facility, these costs can be considerable.

Even more costly, however, and more hidden, are the dumpster fees themselves. A typical dumpster used for this purpose will carry about five tons of wood if it is well organized. The dumpster has to be rented, and there is a fee for pickup as well. This all adds up to a very high cost per ton to get rid of that five tons of waste. Utilities typically only think of the tipping fee when they add up the disposal cost, but the tipping fee may only be 25 percent of the total cost. The waste management company has often buried a lot of hidden costs in there.

Alternatively, a semi-truck picking up the old poles could carry closer to twenty-five tons, give or take. What Cox Recovery does is pick up the old poles when we deliver the new ones, sending them back on the truck rather than sending the truck back empty, thus moving a higher tonnage of poles more cheaply and efficiently. Not only does this reduce costs for the utility, but it also reduces our environmental footprint in terms of greenhouse gas emissions. We take the old poles back to a centralized yard, make all the arrangements, and prep the material for whatever its next stage in the life cycle will be—I will discuss the various options in a later chapter.

This simple idea was already a step forward in innovation; more substantial and disruptive innovations follow in the wake of this, but

Cox Recovery at least gave us a clear path to what those other innovations would end up being. As I describe in the following chapters, this has morphed into the possibility of Cox's entering into the renewable fuel business—which goes to show that innovation should not be seen as a one-off idea, but rather as a fertile ground that a lot of other things can grow from. First, though, it is important to point out how this kind of innovation lends itself to disruption via brand differentiation, which can be hard to come by in commodity industries; this is the subject of the next chapter.

PATH TO SUCCESS

Another component of our aim to offer a turnkey service is that we need to be fundamentally disposal agnostic. We don't want to come in with a one-size-fits-all disposal solution; that's what the waste management companies tend to do, and that's what has gotten us into the situation we're in. Our goal is to present a menu of disposal options for a utility to choose from. They all have different costs and offer different levels of environmental value; we are not biased toward one option or another, so we just go with the option that is locally available (according to state laws) and that offers the best balance of environmental value against a utility's budgetary constraints. We work with the utility to figure out what makes the most sense.

> ***Our goal is to present a menu of disposal options for a utility to choose from.***

Along with our agnosticism about disposal methods, however, comes the realization that none of the options is entirely satisfactory; this has put us on the path of developing yet another method of disposal, one that would be cost effective while also accommodating all types of chemical treatment, including, of course, CCA. Until such a method is developed—and we are currently in the process of developing one—effective and sustainable life cycle management of utility poles will be incomplete.

If we are successful in our enterprise, we could set the new standard not just for how utility poles are disposed of, but even for how they are sold—namely, as having the capability to be turned into a renewable product. Our total product life cycle model ultimately has the potential to disrupt a commodity market—to change the paradigm of the utility pole industry.

Much of our success will depend on the utilities being receptive to our new business model. Much of this, in turn, depends on who comes to the table in the discussion of pole buying from the utility side. If sustainability specialists are involved, for instance, we typically find that the utility will find our model to be of real value to the company and will want to work with us. In the next chapter, I will discuss our project of bringing all the relevant stakeholders to the table in the pole-buying process, and our attempt to make inroads through our emphasis on the bottom-line benefits of the Cox Recovery model of Total Lifecycle Pricing.

CHAPTER 5

Escaping the Commodity Trap

How much time do you spend at the hardware store when you're buying lumber, choosing exactly which boards you want to use? Apart from some minimal due diligence (i.e., don't pick one that's rotten, or broken, or only half as long as it's supposed to be), there probably isn't a lot of picking and choosing, careful comparison of individual boards, and selection of "just the right one"—the kind of care someone might put into purchasing a Christmas tree or Halloween pumpkin, for instance.

One difference between boards you do most likely consider, though, is price. A board is a board, so the cheaper the better, right? That is because the lumber market—like the utility pole market—is a commodity market, meaning the products sold are highly standardized. Brand differentiation is not common in commodity markets; nor is innovation. What *is* common is downward pressure on prices; often, for the consumer, the cost is the only differentiator for commodity producers, so the only competition that occurs is a race to the bottom on price.

Brand differentiation is not common in commodity markets; nor is innovation. What is common is downward pressure on prices

As in any commodity market, the buyer of utility poles—the employee of a utility who controls which poles are purchased from which vendors—has a lot of power to shape the market. If the buyer only looks at price, then competition keeps happening only on the basis of price; this is how industries fall into the commodity trap. Creative, disruptive innovation is needed to help companies escape this trap. This was one of the overarching aims of Cox implementing our innovation process in the first place: to disrupt our commodity market and give consumers (utilities) reasons to buy from Cox that went beyond lowest price.

Creative, disruptive innovation is needed to help companies escape this trap.

The development of Cox Recovery offers a paradigmatic example of innovative methods of escaping the commodity trap: changes in the business model to offer ancillary services and changes in pricing structure. The shifts we have made are disruptive, and potentially transformative, of this market. The next task is to get the buy-in of the customers themselves—the utilities and, specifically, their buyers.

BRAND DIFFERENTIATION

Fundamentally, the market for treated wood products is always going to be a commodity market. If you put two utility poles produced by two forest products manufacturers next to one another, 99.9 percent of the population would not be able to identify any brand-relevant difference between them. That segment of the population includes the people who actually are the customers that companies like Cox want to grab: namely, pole buyers for utilities.

The ways Cox could go about setting ourselves apart in this commodity market mirror the types of strategies used by other companies with the same set of problems. As I've discussed before, product-based innovation—in our case, making a new and better pole—is not a promising avenue for a commodity market. So, one strategy for a company that wants to stand out from its competitors is to shift the innovation focus away from the product and toward services that can be offered *around* the product.

> ***One strategy for a company that wants to stand out from its competitors is to shift the innovation focus away from the product and toward services that can be offered around the product.***

There are precedents for this brand differentiation strategy in other industries. General Electric, for instance, introduced a rental/service model for one of its highest-margin products. The aircraft division found that competitors were forcing prices down on the jet engines they sold to airline companies. While GE still sells engines, they now offer the option of renting engine use by the hour; the airline only pays for the hours the engine is actually being used, and has a contractual arrangement with GE for service on the engine. This helps GE's bottom line while also being attractive to airlines that are looking to reduce the up-front costs of obtaining jet engines.

A more directly related precedent can be found in Cox's own work with radio frequency identification (RFID) for pole inventory and tracking purposes. As I have described, our RFID-based company, SMS, was the product of an attempt to implement an innovation that attached an ancillary service to our utility poles rather than trying to make the product itself better.

This service model is appealing from the perspective of the buyer, which is the aim of the Cox Recovery service model as well. However, the pole buyers for utilities—especially the large, publicly traded utilities—typically do not have sustainability metrics and pole disposal services on their radar. The buyer does not spend a lot of time thinking about where the poles come from, and probably no time at all thinking about where a pole is going to go fifty years down the line at the end of its useful life span. The pole is a line item in their budget; apart from due diligence, they will decide based on price.

We have to blow up that line item, to show that we can provide value to the buyer's company even though we may not have the lowest price point. This falls into two categories that each have to work together: bringing more stakeholders to the pole purchase discussion, and once they are engaged, illustrating the power of what we call Total Lifecycle Pricing.

BREAKING DOWN SILOS

The challenge in showing the pole buyer the value of our disposal services lies in the fact that buyers typically see themselves as intervening at one single point in a broader process, while the process as a whole does not concern them. Most often, buyers want to move a needed utility pole from the supplier's ownership, through purchase, to use by the utility in the field, at the minimum price. They do not look at the process—the life cycle, the supply chain—in its totality, but are incentivized to focus only on the front end. Getting rid of the stuff is asset or investment recovery's problem.

However, the point of purchase has significant implications way down the line. In particular, since CCA is currently the cheapest chemical treatment, CCA-treated poles are the ones that will be most attractive to buyers in many if not most cases; but they are also the poles that today cause the most problems at end of life. The thought process should be familiar to everyone: I may think I'm getting a really good deal on a very cheap car; but if it is very cheaply made, I'm going to get dinged down the road by maintenance costs and having to replace it sooner rather than later, and the long-term cost will be greater than the cost of just investing in a higher quality automobile. This is exactly the thought process we want the utility pole buyer to engage in. From this point of view, yes, CCA may look like the cheapest option now, but the utility will pay the price, literally, down the road—unless new methods for disposing of CCA are developed.

Getting buyers to adopt this mind-set, which is a real paradigm change for them, will be one of the largest challenges our industry will face in the near future. Oftentimes a buyer's only responsibility, from their point of view, is to buy the cheapest pole available that can be delivered on a regular basis. He or she doesn't necessarily worry about the supply chain at any other level. We, however, encourage buyers to take a broader view—in particular, to think about what is going to be done with those purchased poles at the end of their useful life span.

The only way to get the purchasers to understand what they're paying for and to get them really engaged in the supply chain conversation is to walk them through what the costs of disposal actually are, and to request information on certain aspects of their processes so that we can get them to a more accurate understanding of what poles are costing them across the whole life cycle. As a result, we often need someone higher up in the organization to take an interest in the pole-buying process from a total life cycle perspective.

Constructing an effective disposal strategy typically involves multiple constituencies coming together to discuss their needs. We need someone to represent every stage of the life cycle of the pole. Operations personnel help provide guidance on what can or cannot be done in the field and at the yard level to ensure both employee safety and efficiency and reduce a utility's environmental footprint. For example, typically if a utility is able to bring whole (as opposed to cut-up) poles from the field, this eliminates safety issues and often reduces disposal costs by allowing new pole delivery trucks, constructed to handle whole poles, to easily pick up old ones during their travels to utility storeroom facilities. In turn, fewer trips to and from a storeroom to pick up poles helps reduce greenhouse gas emissions. Engaging in similar dialogue with asset recovery and sustainability people at the utility helps to further foster the creation of a well-defined disposal strategy.

Utilities increasingly have strategic entities devoted to thinking about things like renewable energy sources, thus introducing a longer-term perspective on sustainability issues; but a seemingly mundane item like a utility pole may well not be on these people's radars either.

Still, success in finding a sustainable way of disposing used poles will require utilities to start thinking about the pole-buying process strategically and holistically. This means considering disposal-relevant factors—in particular, the form of chemical treatment—at the point of purchase. Sustainability people need to be integrated into the conversation at the front end, so that decisions about buying a certain type of pole are informed by sustainability considerations. Currently, the sustainability professionals only look at the back end: "Can we recycle more?" and so forth. I've never seen a contract for the purchase of new utility poles that had to be signed off on by a sustainability officer, but this is the kind of measure that needs to take place for a meaningful sustainability effort to occur.

I've never seen a contract for the purchase of new utility poles that had to be signed off on by a sustainability officer, but this is the kind of measure that needs to take place for a meaningful sustainability effort to occur.

The pole-buying process, then, needs to involve multiple stakeholders engaging one another, a process likely spearheaded by whoever is in charge of the utility's supply chain. Even the legal department can have a say, as someone there will be able to assess the exposure to liability involved in any particular disposal method. In fact, some of the calls we get from utilities were actually spurred by concerns raised by the legal department.

I've talked at length at industry meetings to utilities and encouraged them to rise above the silos and have some method by which they can have oversight of the entire supply chain, elevating the discussion of both purchase and disposal. This is a forward-looking model. These parties certainly want to buy the cheapest pole, but the decision may be different when they begin considering the costs of end-of-life operations. Disposal, it turns out, is a surprisingly big cost in relation to the price of a new pole—about 60 percent; this usually gets the attention of the heads of the supply chain. Again, we want them to think like the car buyer: Would you want to pay $25,000 for a car if you knew that down the line you would have to spend another $15,000 (i.e., 60 percent of the purchase price) to care for it through the end of its life cycle? What if you could instead pay $30,000 for the same car, but let the car seller take care of those downstream costs?

Disposal, it turns out, is a surprisingly big cost in relation to the price of a new pole—about 60 percent; this usually gets the attention of the heads of the supply chain.

Once you start looking at disposal, it becomes clear that there are opportunities there not just for more environmentally sustainable practices, but for more competitive pricing under a different business model.

TOTAL LIFECYCLE PRICING

With Cox Recovery, we have essentially taken this thought process and revamped our business model around it. In this model, the pricing structure of our utility poles, at the point of purchase of the new pole, incorporates the cost of disposal. Effectively, we are helping utilities capture the true ownership cost of the pole—

In this model, the pricing structure of our utility poles, at the point of purchase of the new pole, incorporates the cost of disposal.

similar to purchasing a service contract for a vehicle at the time of the vehicle's purchase. This is the bottom line of our strategy for incentivizing the customer to think about the project holistically and bring the collective group together for the pole-buying conversation. When these parties get together, they can look at the pricing of the new pole and the disposal service and see that it is in line with the interests of each stakeholder.

This is, again, a paradigm shift for utilities with regard to the bidding process. If a bid includes only the cost of the new pole and not the cost of its ultimate disposal, then of course it is going to have a far lower price and thus be more attractive to the buyer. This means that when we talk to utilities, we have to put a lot of effort into explaining our process, how we arrived at the price that we arrived at, and how, in spite of the seemingly higher price, this will positively impact the company's bottom line when looked at from a different perspective.

Cox Recovery's disposal service is handled as part of the process of contracting for the purchase of new poles. Another result of this—which serves as an additional incentive for utilities to look into our model—concerns the budgeting of the costs involved in pole disposal. In most utilities, disposal is currently considered a part of operations, so the costs are incorporated into the annual operating and maintenance budget. As I mentioned earlier, this can be up to 60 percent of the cost of a new pole, so it can be a significant line item for many utility budgets.

New poles, on the other hand, are considered a capital expense, since they are installed and expected to operate for forty to fifty years. With our Total Lifecycle Pricing method, a utility can subsume the costs of pole disposal into that expense, since we treat it as part of the maintenance of the pole, and thus the disposal cost moves from the

O&M budget to the capital expense budget. This has a couple of effects from a financial point of view.

> ***A utility can subsume the costs of pole disposal into that expense, since we treat it as part of the maintenance of the pole, and thus the disposal cost moves from the O&M budget to the capital expense budget.***

First, the utility can now depreciate the cost of disposal just as they traditionally would the cost of a new pole. With a new pole, the utility can deduct a small amount of money from year to year from the cost of the pole for tax purposes. With disposal built into the price of a new pole, disposal pays off from a depreciation standpoint, essentially providing a small financial incentive.

Second, and perhaps more interesting, utilities are set up so that they get a guaranteed return on all capital investments; this typically happens through rate increases. Utilities have to get approval from local agencies to increase the rates that they charge individual consumers; if they can show that a given rate increase is needed to pay capital expenses, it will likely gain approval. By pushing disposal costs into the capital expense budget, utilities are able to effectively turn disposal from an annual expense to a capital investment with associated financial benefits to the utility.

Typically, somewhere along the way as we lay out this program to our customers, they come to grasp the benefit of focusing more on disposal as part of their purchasing criteria. Here again, innovations in the area of sustainability result in cost saving and increased profitability downstream, not just for Cox but also for the utilities that are our customers. If we can successfully develop a business model and a pricing structure that looks at the total life cycle of a pole, we will be able to change, potentially, the paradigm of how the industry itself looks at buying poles, from the point of initial purchase forward.

We are confident of the market response—while it may take some work to change traditional mind-sets, and while it requires some behavioral change from those in the utility industry more broadly, there is an eagerness for a new model here among utilities that are thinking strategically. Also, while we initially were attracted to the synergy in logistics between our existing business model and the area of pole disposal, this innovation has been paradigm changing for *our* business model as well, changing the way we approach the sale of new utility poles.

By pushing disposal costs into the capital expense budget, utilities are able to effectively turn disposal from an annual expense to a capital investment with associated financial benefits to the utility.

CHAPTER 6

Disposal Options Today

For many years, the Shenandoah Valley Electric Cooperative, a Virginia-based utility located inside the confines of Shenandoah National Park, disposed of their used poles by having a local waste hauler take them to a nearby landfill. But one day Cox Recovery received what might best be termed an emergency call from the co-op. The local landfill had run out of space; the co-op was told they needed to find a new home for their used poles, and suddenly the utility needed help in identifying an alternative disposal option. Shenandoah Valley ended up becoming one of Cox Recovery's first customers.

This utility's response to a disposal dilemma reflects what we see regularly across the country. First, few utilities have a defined long-term strategy for wood disposal, and most have instead become reliant on what has historically worked best and cost least for them: landfilling. Second, due to the lack of a disposal strategy, few utilities have a good understanding of what other disposal options may exist for them. The initial aim of Cox Recovery is to meet the demand created by this situation for a way for utilities to connect with more sustainable disposal solutions.

Of course, an integral part of the disposal strategy is determining which disposal option or options best fit with a utility's needs. As mentioned, state law will, to a degree, dictate which options are available, in addition to the proximity of these options to a given utility's service area. Generally speaking, all involve a trade-off between affordability and level of environmental stewardship. On the whole, the better an option is with regard to sustainability, the more obstacles and costs there are to a utility participating or connecting with that solution. In this chapter, I discuss each option, along with its pros and cons from both environmental and cost standpoints.

REPURPOSING

The first, most straightforward method is what we might call simple reuse or repurposing. In the past, going back decades, this actually was a much more common way of disposing of old poles. In rural areas in particular, a utility would simply chop a pole up into manageably sized pieces and then sell them, give them away, or even just leave them out for Farmer Fred to pick up and use as a fence, build a retaining wall, or serve myriad other needs.

The upside of this approach for utilities is that it is low cost, or even in some instances profit generating. However, it has considerable downsides, so that it has become a much less widely used disposal method in recent decades. Those who still engage in the practice do so in a much more limited and cautious way than in the past.

The key downside, particularly in today's legal environment, concerns the potential liabilities that a utility may assume in implementing this type of effort. Utilities might still leave wood out for Farmer Fred to build a fence out of, but they will almost surely make him sign a release form to mitigate their liability. After all, this wood is chemically treated, and utilities could find themselves liable if the wood gets used in a way that causes some harm. There have been instances where people took treated utility poles and mistakenly burned them as fuel for a wood-burning oven to heat their houses; they developed cancer down the line due to incinerating chemicals that become carcinogenic when burned, and they sued the utility for it. Utilities, of course, want to avoid this type of scenario; they've all heard the same horror stories, and unless they have a very high risk tolerance, many just say, "We don't even want to mess with that."

Another challenge with attempting to build a disposal program around repurposing poles is that only a relatively small amount of the wood collected is actually in good enough condition to be used again. Typically, only about 40 percent of the material to be disposed of is really reusable anyway, and the rest—whether it is rotted out, full of nails, or whatever—is landfilled. Sometimes utilities may believe they are headed down a more sustainable path by implementing a disposal program based on repurposing poles, when in fact 60 percent of that material is not marketable as a reusable product, so it goes to a landfill either way.

Repurposing, then, is at best an incomplete solution; it might work in tandem with another approach, but if a utility is pursuing this as its only alternative to landfill, then the environmental value they are producing at the end of the product's life cycle is minimal. Eventually, even the fence post that was previously a piece of a utility pole will no longer be usable and will have to come down, and where will it go then? Most likely the landfill. Repurposing in this form does not so much give the product a second life as it just extends the first one, passing the responsibility for disposal off to Farmer Fred when he's finished using it.

LANDFILL-TO-GAS

One relatively low-cost, semi-alternative to landfill is to deploy the material housed in the landfill itself as an alternative energy source. As landfill operators have started to become tapped out in terms of the space they have to offer, many have invested in accessing an alternative revenue stream by converting their landfills into plants that produce energy by taking in the methane gas coming off of the landfilled material. As the landfilled material decomposes, it produces methane, and these plants capture that gas underground and transport it via tubes to the surface.[6] The continuous flow of methane powers turbines that allow each of these landfills to produce, on average, about six megawatts of electricity annually.

While landfill-to-gas (LFG) is a relatively new development, the EPA has thrown its weight behind this option, since it turns already existing landfill into a source of at least some value and reduces the greenhouse gas emissions of the landfill; over four hundred such facilities are now operating nationwide, with more in development.[7] It also benefits the landfill owner, of course, who may be losing revenue because the landfill is running out of space, by providing a

---

6 Kate Ascher and Frank O'Connell, "From Garbage to Energy at Fresh Kills," *The New York Times*, last modified September 15, 2013, https://archive.nytimes.com/www.nytimes.com/interactive/2013/09/15/nyregion/from-garbage-to-energy-at-fresh-kills.html.

7 "Landfill Methane Outreach Program (LMOP)," EPA, accessed June 15, 2018, https://www.epa.gov/lmop.

new revenue stream. It is attractive to some utilities that are trying to comply with alternative energy mandates.

Landfill-to-gas may be the best environmental solution at the moment for material that by state or federal law must be landfilled, such as CCA-treated poles. However, from a more forward-looking perspective, the prospect of continuing to fill landfills is not a satisfactory option.

Landfill-to-gas may be the best environmental solution at the moment for material that by state or federal law must be landfilled, such as CCA-treated poles.

WASTE-TO-ENERGY

A waste-to-energy (WTE) facility takes raw material waste and, as the name suggests, converts it to energy by incinerating it at a high temperature. This either generates heat or electricity or provides boiler fuel to generate steam or gas energy. Most of the raw material for these facilities comes in the form of municipal solid waste—in other words, garbage. Typically, a facility will contract with a municipality to collect its garbage to fuel the plant. However, many facilities will also take ground-up treated wood as fuel. A facility that takes wood will work similarly to a landfill, charging a fee, analogous to a landfill's tipping fee, to accept and dispose of the material.

From an environmental point of view, this option is much better than the previous two. The emissions from WTE facilities are strictly regulated through the use of scrubbers, and a quality facility will take one ton of waste and produce one megawatt of electricity, enough to power over three hundred homes for an hour.

The availability of this option varies from state to state based on varying laws regarding whether or not treated wood can be incinerated. Even in states where it is allowed, facilities typically can only burn treated wood as a small percentage of what they use in municipal solid waste. Utility poles are certainly not going to be a WTE facility's primary source of fuel. Also, again, this can only be done with penta- or creosote-treated wood, since CCA-treated wood cannot be burned. The economy-minded utility that opts for buying cheaper, CCA-treated new poles, then, does not have this option.

I mentioned, also, that with these disposal options there is always a trade-off between environmental stewardship and economical efficiency; and the improved environmental value of a WTE-based disposal strategy comes with an added cost. A utility that does opt for the WTE option not only has to pay the facility's fee, but also faces the challenge of preparing the wood for the facility—by grinding it up. Any non-wood material—nails, staples, attachments of any sort—has to be removed. This means extra process steps, extra operational costs, and extra logistics. At the end of the day, consequently, many utilities will look at this option and decide it is just too costly.

Many utilities will look at this option and decide it is just too costly.

This option is also geographically restrictive, since any facility has to be state approved, meet applicable emission standards, and be permitted to take treated wood as a fuel source. Most WTE facilities that will accept treated wood are in the Northeast or Midwest. Logistics, and transportation in particular, is a major part of the cost structure for utilities considering these disposal options, so a utility will have to determine how far it is willing to transport material to make accessing a WTE facility worthwhile.

BIOMASS

Similar in ways to WTE, biomass energy production involves incinerating an organic material to fuel boilers to provide power for very energy-intensive industries, such as the manufacture of cement. It largely shares the environmental benefits of WTE regarding the energy captured, and thus the value generated, from the waste. It also has the upside of being more desirable to the utility from a cost standpoint: instead of having to pay a facility (whether landfill or WTE facility) to take the waste wood, biomass facilities are actually much more reliant on this type of feedstock as a fuel source, and will thus pay suppliers for the material. A cement company, for example, may use this waste stream as its primary energy source, unlike a municipal WTE facility that has to limit the amount of waste wood it uses.

While a biomass-based disposal strategy offers potential environmental benefits compared to other options, it also comes with added risks and costs. First, the biomass facility typically contracts with the supplier (the utility) for a steady supply of a specified amount of product of a certain quality. The commitment to a certain volume provided at a steady rate can be a challenge for a lot of utilities, which either do not produce enough waste to be worth the facility's while or do not do so consistently—in the North, for instance, some utilities will shut down their pole replacement operations entirely in the winter, so there would be little possibility of supplying a facility with waste wood year-round.

The biomass option comes with many of the same extra operational costs as WTE, with the added wrinkle that biomass facilities are a good bit more selective regarding the quality of the chipped wood product itself. If the wood pieces are not within a certain size range, the biomass facility will not accept it. Also, to get the relationship started, the supplier has to provide samples so that the boiler operator can make sure it meets their standards, testing the moisture content to determine the temperature at which the wood will burn. After that, the supplier is expected to provide material within the range of a certain moisture content or face a reduction in the income provided to them.

The operational costs added by the need to meet these specifications, in addition to potential financial penalties for failure to meet contracted standards or levels of volume, mean that biomass is likely only cost effective for very large producers of waste wood—for instance, railroad tie disposal companies, which dispose of a huge amount of discarded treated railroad ties on an annual basis.

> ***Biomass is likely only cost effective for very large producers of waste wood.***

There is also currently a great deal of uncertainty in the biomass route for a couple of reasons. The first regards the market: with the relatively recent price drop in natural gas—which is an alternative to biomass—and the oversupply of biomass in many parts of the country, the market price for biomass has declined precipitously. In fact, many large biomass providers are now sitting on large quantities of material in hopes of a price rebound.

The second arena of uncertainty has to do with changes in federal law. There is controversy over what materials are acceptable for use as biomass. Creosote-treated poles are generally considered acceptable, but debate is happening at several levels over whether penta-treated poles can retain their status as an EPA-approved biomass material. While penta-treated poles have been declared nonhazardous, environmental groups continue to lobby against the use of this material in either a WTE or biomass facility. The uncertainty is replicated at the state level, and it changes from one administration to the next, so that ultimately biomass has much of the same geographical restrictiveness that WTE has.

LOOKING AT DISPOSAL OPTIONS THROUGH THE LENS OF INNOVATION

Examining these various options as part of a broader disposal strategy suggests two avenues of further innovation, both of which Cox has stepped in to pursue since launching Cox Recovery. First, the highest value-added options available today (WTE and biomass) are currently also the costliest for utilities to access or participate in. A service provider that eased this access and helped offset these costs could potentially meet a major market demand and generate much greater environmental value for utilities.

Second, while WTE and biomass are currently the best environmental options for disposal, they remain woefully incomplete in that they can only process penta- and creosote-treated poles, when in the coming years, as I have discussed, CCA-treated poles will make up an ever-increasing portion of the wood waste stream. A more satisfactory disposal solution needs to come along to accommodate these poles.

CHAPTER 7

Moving to the Circular Model

The current slate of disposal options leaves us with a tough situation; in every case, we run into a disposal situation where there are significant costs (financial or environmental) and obstacles to implementation. From a life cycle perspective, we keep running into a wall at the end of the process, which moves from development of the raw material (tree growth), manufacture (cutting and treatment), use, and finally disposal. You grow the tree, you harvest the tree, you manufacture it, it goes up in service, and then ultimately it gets disposed of one way or another.

It's that last step, the end of the line, that keeps presenting difficulties. Even in the best-case scenarios currently available, the material has to somehow *go away*, albeit in a relatively environmentally friendly way that can produce some value in the form of energy. Even reuse, as I mentioned, does not keep the pole from eventually being a waste product. It may extend its life, but eventually disposal will become an issue anyway. It's still the same product. From the utility's perspective, the problem remains one of getting some third party to get rid of it for them.

These types of issues are an inherent part of an economic model where the product is discarded at the end of its life—what is now known as the "linear economy." For centuries, it was about the only model anyone knew, and it has left us with the problems we have today: what to do with all of this *waste*, much of which is damaging our atmosphere. While the problems with this model have been mitigated by improved production practices and by the rise of recycling, the business model of the vast majority of companies still takes the linear model for granted.

Given the dominance of the linear model, breaking out of it represents a hugely disruptive form of innovation. The current limitations on disposal options were pushing Cox Recovery in this direction, which led us to the alternative model that has developed recently under the heading of the "circular economy" (CE). In this chapter, I discuss the ideas behind the circular economy, and in the next chapter I will explain Cox Recovery's own attempt to break into this model via the use of waste wood to create a synthetic diesel product.

Business is about making money over the course of months and quarters, while sustainability often requires a substantial short-term cost for the sake of a long-term benefit.

THE BASICS OF THE CIRCULAR ECONOMY

Many businesspeople think that while environmental sustainability may be a noble and worthwhile goal, it is still inherently in tension with business. Business is about making money over the course of months and quarters, while sustainability often requires a substantial short-term cost for the sake of a long-term benefit that is not easily thought of in business terms. Still, engaging with sustainability has become unavoidable, and the circular economy has emerged in recent years as an approach that allows businesses to pursue greater environmental sustainability in a way that also leads to greater value for the company.

The basic idea of the circular economy is rooted in the type of product life cycle thinking I described in Chapter 2. The linear model of a product's life can literally be represented by a line:

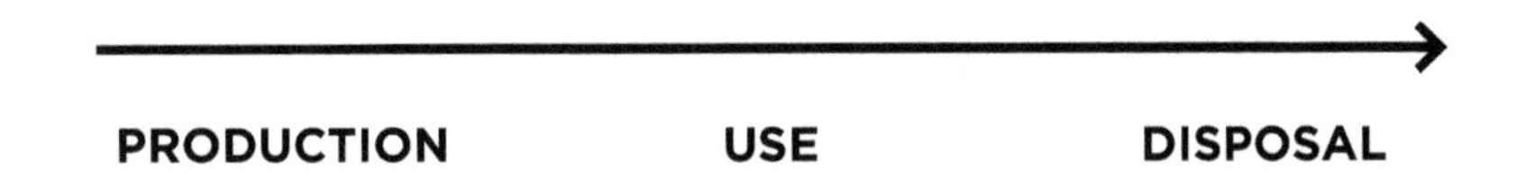

PRODUCTION USE DISPOSAL

The idea behind the circular economy is to take that line and turn it into a circle by looping the end back around to the beginning, or at least to an earlier point along the line. If we view the product life cycle as a stream flowing from design, through production and use, to disposal, the intervention of the circular economy is to divert the stream at the point of disposal and cycle back to a point upstream. In other words, rather than just treating the remaining raw material at the end of the process as waste to be gotten rid of, companies have the opportunity to innovate by closing the loop and turning that material into input at an earlier phase of the process.[8] This can include the production of a new product or even the production of the product that is being disposed of itself.

8 For a comprehensive resource on the circular economy idea, see: "What is a circular economy?," Ellen MacArthur Foundation, accessed June 15, 2018, https://www.ellenmacarthurfoundation.org/circular-economy.

Product life cycle management, or looking at supply chain and value chain from a holistic perspective, is clearly a first step on the way to a circular model. For a company to be able to divert its used products from disposal (which is so often going to be in a landfill) back into the production process, it has to have some control over the entire product life cycle—not only how it's conceived from a materials standpoint and how it's designed and manufactured, but also its use and disposal. This means treating the raw material of your product as an asset to be taken advantage of *beyond* its being an input to the initial production process. From this point of view, letting the raw material be simply discarded at the end of its useful life starts to look like a mistake from a bottom-line point of view—a waste of a potential resource.

> ***Letting the raw material be simply discarded at the end of its useful life starts to look like a mistake from a bottom-line point of view—a waste of a potential resource.***

The question then is how you can take the raw material output and loop it back around to an earlier stage in the production process, so that the process becomes regenerative rather than linear. This often involves shifting the company perspective on its products so that they become assets to be deployed as a service to consumers or end users; this "product as service" model is widely discussed in the CE literature. In this model, instead of the consumer purchasing the product and taking full ownership of it, the vendor simply provides the product for the time of its use and retains ownership, so that it can take the product back at the end of that time and find a second use or second life for it.

The CE model is a value generator from both the business and the environmental points of view.[9] The heart of the CE philosophy is the ability to reduce the environmental impact of your business through the reuse of what would typically be considered normal waste by-products. The circular economy is a relatively new trend, but already many businesses are taking on the model for purposes of both environmental sustainability and profitability—both achieved through the elimination of waste. Environmental constraints suggest that businesses ought to incorporate the idea of the circular economy into their business models. Again, this is an area that is ripe for innovation.

This also makes business sense, as the elimination of waste is a cost-saving measure—landfills, for instance, are getting more and more expensive to use—and it can be a brand differentiator and create deeper customer relationships, the kind of "stickiness" that makes customers or clients adhere to a particular vendor. Repurposing the disposed-of product in a way that it can somehow be led back to the customer is a powerful model. This often requires a change in business model. A 2013 article on the *Fast Company* website outlines five business models that adhere to this basic notion of the circular economy.[10]

A company that adopts a CE approach focuses mainly on a couple of things. The first is building a more environmentally friendly business model. The second is making the company more profitable; for circular endeavors to gain traction in the long term, they ulti-

---

9 "Why the circular economy is all about retaining value," McKinsey & Company, October 2016, https://www.mckinsey.com/business-functions/sustainability-and-resource-productivity/our-insights/why-the-circular-economy-is-all-about-retaining-value.

10 Peter Lacy et al., "5 Business Models That Are Driving The Circular Economy," *Fast Company*, last modified April 24, 2013, https://www.fastcompany.com/1681904/5-business-models-that-are-driving-the-circular-economy.

mately have to contribute to the business's bottom line. Fortunately, insofar as circularity is a way of eliminating waste and taking advantage of access to raw material assets, these two objectives can go hand in hand. Often companies look at circular economy or even total life cycle management and see only the up-front costs, the higher cost structure to their product, and the operational challenge of shifting their business model.

> ***For circular endeavors to gain traction in the long term, they ultimately have to contribute to the business's bottom line.***

The circular economy is clearly an area where innovation becomes disruptive and transformative of business models, rather than simply making incremental improvements of existing practices.

WHOSE IDEA WAS THIS?

The circular economy idea is the convergence of a lot of lines of thinking about manufacturing, design, logistics, and business models. One of the most long-standing, and most familiar, exemplars of the circular approach can be found in the realm of packaging, which results in a waste product, whether paper or plastic, that can then be recycled into new packaging. This creates more value environmentally than filling landfills with packaging or destroying the waste to produce energy. The ingredients of the original product are actually put back into the new product.

Closing the loop in this way has become a reliable strategy for facing regulatory and market demands on waste reduction. One industry that has met some of the same challenges we are seeing with waste wood is the carpet industry. Landfills are filled with massive amounts of old nylon carpeting, and government mandates are starting to force carpet manufacturers to divert large portions—in some cases all—of their waste from the landfill. The companies that have been most successful in meeting this challenge have done so by developing systems for both retrieving old carpet from customers and converting the nylon in that material into usable new product.

The carpet company Desso, in particular, has been a pioneer in this area, adopting a cradle-to-cradle, circular approach, employing a combination of recyclable yarn and consumer take-back programs to use the same material again and again to make their product.[11] Their model is also similar to our own in that they are pursuing this initiative primarily in their business-to-business transactions; businesses in general are usually more amenable to these sorts of take-back programs than individual consumers.

There is in fact an entire 501(c)(3) built around diverting carpet from the landfill. Carpet America Recovery Effort (CARE) focuses on market-based solutions to this problem.[12] The extensiveness of this program is impressive; the carpet industry has had to react in a broad-based way to the demands for waste reduction. There have certainly been challenges, but the carpet industry provides a useful example of an industry that has already faced the same problems we will be facing in the coming years. This example illustrates how companies that are innovative with the CE model will be the ones that end up with a leg up in the long term.

The economic and environmental benefits of the CE model can also be seen in incentivized return and refurbishment programs in

---

11 "Cradle to Cradle design of carpets," Ellen MacArthur Foundation, accessed June 15, 2018, https://www.ellenmacarthurfoundation.org/case-studies/cradle-to-cradle-design-of-carpets.

12 Carpet America Recovery Effort, https://carpetrecovery.org.

electronics and related materials.[13] Hewlett Packard has been a pioneer in this area, diverting millions of ink cartridges with its take-back and recycling program.[14] The company started the program in 1995, and by 2000 they were closing the loop by manufacturing new ink cartridges out of this old plastic. Their "Instant Ink" subscription service automates this take-back process—when one ink cartridge starts running low, the company sends the user a new one along with a postage-paid envelope to send back the old one. HP reports similar difficulties regarding local laws and regulations that tend to emerge for attempts at implementing the circular economy. In their case, for example, the EU classifies all electronics as waste in a way that makes it considerably more difficult to carry that material across borders.

Instant Ink represents the shift from selling products to selling services, as well. We see this shift also in GE's program of renting jet engines out to airliners rather than selling them, as well as in Philips Lighting's contract with the Washington Metropolitan Area Transit Authority (WMATA) to provide LED lighting for twenty-five parking garages.[15] The system that Philips installed provides the lighting when and where it is needed, while Philips itself retains

13 "Establishing a reverse supply chain for electronics," Ellen MacArthur Foundation, accessed June 15, 2018, https://www.ellenmacarthurfoundation.org/case-studies/establishing-a-reverse-supply-chain-for-electronics; "Collection, refurbishment and resale of mobile phone handsets," Ellen MacArthur Foundation, accessed June 15, 2018, https://www.ellenmacarthurfoundation.org/case-studies/collection-refurbishment-and-resale-of-mobile-phone-handsets.

14 Coro Strandberg, "HP and the Circular Economy," HP, last modified 2017, http://h20195.www2.hp.com/V2/getpdf.aspx/c05364027.pdf; Judy Glazer, "Subscriptions for Printer Ink? It's Business for the Future," U.S. Chamber of Commerce Foundation, last modified January 12, 2016, https://www.uschamberfoundation.org/blog/post/subscriptions-printer-ink-its-business-future.

15 Kristine Kalaijian, "Philips' Refurbishment Plans: Cats Aren't the Only Ones with Nine Lives," U.S. Chamber of Commerce Foundation, last modified December 17, 2015, https://www.uschamberfoundation.org/blog/post/philips-refurbishment-plans-cats-arent-only-ones-nine-lives.

ownership of the system infrastructure, cutting costs for the WMATA and leading to energy savings for both organizations.

Other examples involve the use of waste as a resource for a company's internal operations. A high-profile example is the Walt Disney World Resort, where all food waste from its restaurants, including cooking oils and grease, is sent "to a nearby 5.4 MW anaerobic digestion facility owned and operated by Harvest Power. The organic waste is converted into renewable biogas (a combination of carbon dioxide and methane) to generate electricity, with the remaining solid material processed into fertilizer. The energy generated is used to help power central Florida, including Walt Disney Resort's hotels and theme parks."[16] Here we have waste as a feedstock for energy production, as in a waste-to-energy facility, but one where the energy is cycled back to power the business that produced the waste in the first place.

NEW MODEL, HERE TO STAY

Waste Management is a company that ran up against some of the same problems that motivated Cox Recovery. Specifically, their business model, which has historically relied heavily on landfills, was under threat due to dwindling landfill space.[17] In response, they have moved heavily into the circular economy, not only with recycling but

16 "Achieving a Circular Economy: How the Private Sector Is Reimagining the Future of Business," U.S. Chamber of Commerce Foundation, accessed June 15, 2018, https://www.uschamberfoundation.org/sites/default/files/Circular%20Economy%20Web%2011.11.pdf; Marc Gunther, "Disney World's biogas facility: a model for converting food waste into energy," *The Guardian*, last modified October 17, 2014, https://www.theguardian.com/sustainable-business/2014/oct/17/disney-world-biogas-food-waste-energy-clean-tech.

17 Carolyn Pincus and Kate Ellman, "Philips Lighting, WM transition to the circular economy," GreenBiz, last modified December 1, 2017, https://www.greenbiz.com/article/philips-wm-transition-circular-economy.

also in offering sustainability consulting services to the businesses that contract with them, partnering to help develop longer-lasting and CE-friendly products and services.

This example, like the ones that preceded it, show that CE is not just a fad, nor should it be seen as a cumbersome imposition on existing models that should only be taken on to satisfy environmental regulators. Instead, it is a value generator and a potential profit center, one that companies need to be more proactive about participating in if they are to meet the demands of the new economy that is developing.

Circular economy thinking was a big shift for Cox, too. Sometimes innovation efforts that start with a targeted focus like disposal have unintended, broader strategic consequences down the road. Cox Recovery was a rather simple step forward for us in entering the pole disposal market. What we didn't anticipate was the potentially much bigger transformational journey that lies ahead for both our pole disposal efforts and, ultimately, our company business model.

CHAPTER 8

Cradle to Cradle

With the circular economy (CE) concepts in mind, Cox Recovery was able to come full circle, so to speak, back to its primary aim: finding a method of disposing of wood waste that would maximize both environmental value and value for the utilities that are its customers. One common characteristic of the products that lend themselves to the CE model is that they are organic, so they can simply be cycled back into the biological processes involved in the production process.

Wood waste, of course, fits into this organic category, which suggested to us that a more value-generative, circular innovation of some sort ought to be possible. Furthermore, such a solution—providing a disposal option other than landfill for CCA-treated poles—would have a tremendous long-term benefit. For that reason, we wanted to go beyond the current services that Cox Recovery was offering, and beyond the already available disposal options.

We ultimately want to arrive at a mentality of not just providing disposal solutions to utilities, but also adding environmental value through a circular methodology, so that, at the end of "useful life" in one form, we can give the product a second useful life in another form. We are well placed to do this, because we have control of a large portion of the product life cycle. As a result, closing the loop is within our control. The circular economy often requires a change in business model, and this is certainly what we have been looking into—not to replace our old one as much as to supplement it with new avenues of opportunity.

The challenge for us, from a circular perspective, was to figure out how to take a used chemically treated pole and somehow, ultimately, develop a method for capturing some of the value of that pole and channeling it back into the production of a new pole. This, then, becomes a question of innovation.

ENTRY TO A NEW INDUSTRY

So, what could we do with this material? The main aim was still to keep it out of the landfill. In the process of thinking this through, we were interacting with a lot of waste-to-energy and biomass firms, and by extension we were introduced into the world of alternative fuel sources. As I've mentioned, we don't have the resources to establish a separate R&D function, to fund a lab with people in white coats walking around with beakers doing chemistry experiments. Because of this, we have learned quickly that relationships with third parties who do that kind of work are necessary if you're going to get anything accomplished in this direction. In many cases, these alternative fuels did not necessarily require the material to be incinerated the way waste-to-energy or biomass would. Rather, the material would be heated to a point where its structure would start to break down at the molecular level. At this point we had our "aha" moment: if waste wood could be used as a feedstock for this process, the organic material could be broken down into fuel without being burned, which meant it could be used no matter what chemical it had been treated with.

If waste wood could be used as a feedstock for this process, the organic material could be broken down into fuel without being burned.

We also started discussion with a number of other companies that are entirely focused on converting waste products into some kind of fuel (primarily ethanol). We had a lot of conversations with people in this realm, and I began asking, "If we could give you X tons of treated utility poles, could that material be used as a feedstock to produce your alternative fuel? If so, would you be interested in investing with us to transport the material to you?"

The answer to the first question, it turned out, was "yes," but only a few of these companies were really interested in working with us. Also, there were considerable obstacles to our partnering with most of these firms. First, they tended to be located in areas far enough away that it would have been prohibitively expensive to transport material to them—once again, geographic limitations play a key role. The second difficulty was that their production requirements, like those of a biomass facility, called for such a large amount and steady supply of material that we might not have enough waste to deliver to get the process going.

One company we engaged with, Enerkem, was a great fit in terms of mission; they take waste, including wood waste, and turn it into ethanol fuel. However, they were building plants entirely in Canada, and the scale of their operations was so huge that it would have been a major challenge to get material to them in a way that was at all cost effective—we really would not have been able to produce enough waste to feed their machine. To give a sense of the scale of their facility: They have since partnered with the city of Edmonton, Alberta. They take the city's entire supply of municipal waste (garbage) and turn it into ethanol that is pure enough to be sold at the gas station.

The situation was beginning to become discouraging when all of a sudden a newsletter for the biofuels market, *Biofuels Digest,* broadcast an open call for people working in this area to be interviewed for upcoming editions. I thought, "Why not?" I did an interview, and the day after it appeared in the newsletter, I received a call from a reader in Alabama, Paul Stabler, who was working with a company called Proton Power. He said, "I saw that you're interested in trying to work with treated wood. I think we might have an opportunity to help you."

I told him I was interested, but I also shared some of the obstacles and frustrations I had run into. He told me that his company could scale their plant to coincide with the needs of the feedstock supplier rather than offering a massive, one-size-fits-all plant that we would never be able to fully supply. It occurred to me then that we might have a major opportunity here; if that kind scaling can be done, then what if we could just build our own plant? This would give us a way around some of the obstacles we kept running into. It was in our conversations with Proton Power that our next innovation opportunity really started clicking into place.

We speculated that an opportunity might be available along these lines: Instead of trying to negotiate with WTE or biomass facilities, why don't we just process the material ourselves? Then we could take this raw material we were picking up in trucks, grind it into chips, and use it as a feedstock to actually produce our own fuel, which we could use for our own internal purposes—mainly, to act as a carrier in the chemical treatment of new poles. Recall that penta-treated poles are treated using a petroleum-based carrier, which makes penta more expensive than CCA, which uses water. Producing our own fuel, though, would make the more environmentally friendly penta much more affordable.

> ***We could take this raw material we were picking up in trucks, grind it into chips, and use it as a feedstock to actually produce our own fuel.***

Cox consumes several million gallons of diesel fuel a year just in treating utility poles in our manufacturing process, and this becomes expensive; so if we could produce all of that fuel at a lower price and in a more environmentally friendly way, this would be a huge advantage. We would, in a circular manner, be looping the end product, the waste, back around to the treatment stage of the production process, both eliminating (or rather finding a use for) the waste and eliminating our reliance on outside fuel sources. There is no doubt that it would be meeting a major need internally, improving our business

> ***In an ideal scenario, we would be able to switch out traditional diesel with a new synthetic diesel altogether, creating substantial cost savings in our operations. We would be in control of our own destiny.***

performance by lowering the internal cost we pay for fuel and mitigating our exposure to market fluctuations in fuel price. In an ideal scenario, we would be able to switch out traditional diesel with a new synthetic diesel altogether, creating substantial cost savings in our operations. We would be in control of our own destiny.

We would just have to have our own plant for processing the waste into fuel. We began looking into the possibility of partnering with a third party who would own and operate the facility that would turn our feedstock into fuel. This required a good bit of spontaneous market research, as this was uncharted territory for Cox, as we tried to understand how this field of alternative fuel operated from a business perspective.

HOW DOES IT WORK?

Synthetic diesel producers like Proton Power produce a synthetic diesel through a cellulose-to-hydrogen power (CHyP) process, which is what we would need for the treatment of wood. With their modular and scalable plant design, Proton Power could design and manufacture the actual facility for us. As for geographical restrictions, we could in principle put the plant wherever we wanted, wherever would make the most sense for us.

The CHyP process requires Cox Recovery to first process the wood retrieved from utilities for use as a feedstock. This means taking the whole pole, making sure to remove any metal attachments, and grinding it into chips. These chips are much smaller and finer than what we typically think of as wood chips, but this size is what is required for the CHyP feedstock. Given that most of this wood will be chemically treated, our grinding process has to involve strict dust control and other environmental control standards to prevent the release of toxins into the atmosphere.

These fine wood chips are fed into the CHyP unit, where they are heated at a very high temperature (in the range of two thousand degrees Fahrenheit). This converts the solid into a gas. The chemical volatilizes at that point, meaning it turns into a vapor that will ultimately be part of the resulting diesel product. With penta and creosote, the chemical basically doesn't exist anymore. With CCA, any remaining heavy metals will be captured in ash, which is a small quantity of the output of the process; this gets sequestered, encapsulated, and disposed of in another way. Even if this has to be sent to a landfill, it is a massive reduction of material going to landfill. Again, while we are still working on perfecting the process, the goal is to ultimately have a chemical-agnostic process—whatever treatment you have on your poles, we can use them.

This gas is converted into a liquid; that liquid, essentially, is the synthetic diesel, which we have successfully tested as a carrier for chemical treatments of new poles. These treatments effectively close the loop on our product life cycle, providing a cradle-to-cradle solution, with benefits both from an internal financial standpoint and from the perspective of creating environmental value and keeping waste out of landfills.

This type of model could help us set the standard not only for how poles are disposed of but even for how they are sold—as having the capability to be turned into a renewable product rather than being thrown into a landfill. Reducing or even elim-

> ***Reducing or even eliminating the reliance on landfills represents a major advantage and operational cost savings for the utilities we work with.***

inating the reliance on landfills represents a major advantage and operational cost savings for the utilities we work with. This model also leads to a reduction in greenhouse gas emissions, both due to the operational limitations on trucking discussed in previous chapters and from the perspective of diverting material from the landfill: As landfills grow, the amount of methane gas they release as the material decomposes grows correspondingly. Diverting material from the landfill is thus a clear environmental good.

Another beneficial, and potentially profitable, by-product of the CHyP process is what is known as "biochar." Biochar is a substance that contains high concentrations of carbon—which stays in the biochar and thus does not get released into the atmosphere—and can serve as a powerful soil amendment (meaning that when it is added to soil, it can radically increase the productivity of crops, especially in drought-prone areas). The difference I've seen at test farms between crops grown with and without biochar is amazing. It is believed that its use as a soil amendment goes back hundreds of years to indigenous Amazonian people, and currently it is being investigated as a possible avenue of carbon sequestration.

Biochar is a potential revenue generator, as it has a lot of value in the marketplace. Also, from the standpoint of total life cycle management, it is a tool that we can put in the hands of our tree growers to increase the growth rate of trees and make a contribution to reducing greenhouse gas emissions. We are working on how best to take advantage of this by-product of our process—yet another opportunity for innovation.

DEVELOPMENTS ON THE HORIZON

To take advantage of all of the opportunities and benefits that employing the CHyP process offers, the liquid carrier—the fuel—has to meet applicable standards. The American Wood Products Association (AWPA)—the trade organization mentioned earlier, composed of people who treat wood (like ourselves), some academics, people who supply trees to the wood industry, and the utilities themselves—sets complex standards for what is and is not acceptable for chemical carriers. The goal is to create a standard among all treaters for how to treat wood so that there is a general consistency in product quality across the market.

Much of the testing we have been doing up to this point has been directed toward making sure we're in sync with these standards, that we end up with a product that will meet the specifications that the industry requires for it to be considered a viable candidate for use in the treatment process. The type of carrier we're working on is what the AWPA would consider a hydrocarbon solution, and they have a technical committee that is devoted to—and the experts in—this type of carrier. They outline the specific makeup and performance qualifications for carriers for—in our case—penta treatment, and they have their own lab that evaluates whether a given manufacturer's diesel product meets these qualifications (flash point, viscosity, etc.).

We are currently doing internal testing in our own treatment cylinders at our plants. We started testing with untreated wood turned into fuel, just to make sure the resulting fuel didn't gunk up our cylinders or otherwise present problems, and this testing turned out well. We have since done multiple rounds of both laboratory and plant-level testing with favorable results. We are making great strides in refining the quality of the product, which performs as well as or better than regular diesel fuel for our internal chemical treatment purposes, and we are confident about being able to replace our own internal diesel usage supply with this new synthetic diesel in a cost-effective way. We still have several rounds of tests to go through, but on the basis of chemists' predictions, we expect that the results will continue to be positive.

If our tests continue to bear fruit, we plan to build a plant to produce our own synthetic diesel. We are still working on figuring out the ideal location and size of a full-scale facility that could process this amount of waste wood. We expect to start with a facility that can produce about one million gallons of fuel annually, but its modular design will allow us to quickly scale up to a capacity of ten million gallons a year.

As for location, a lot of logistical factors come into play in determining the best option. In South Carolina, Cox is located in what we call the "timber basket"—that is, the spot where our supply of raw material, southern yellow pine, is most plentiful. The correlate for the disposal business—the "waste basket," if you will—would be the northeastern United States. There is a confluence of factors in this area that make it an attractive potential site for a plant. Remember that this is where the U.S. infrastructure boom first took hold, so it has the highest concentration of poles that have been standing for the longest stretches of time. This is also the area where landfill space is most limited, and thus where tipping fees are highest. Collectively, this makes a facility offering a solution such as ours attractive for both utilities and government actors in this area.

We would need to be close to the waste basket to make this operation work cost effectively and from the point of view of logistics, at least initially. Not only would the Northeast provide most of our feedstock in terms of sheer volume, but most of the poles in that area are penta treated; CCA becomes more and more common as you move south. This means less investment up front in the extra environmental controls required to deal with CCA. We also have to deal with the differences between states, some of which will potentially provide incentives for bringing synthetic diesel production to them.

There is another potential avenue for innovation in these developments. If the fuel is pure enough and meets the requisite standards, we can potentially bring that to market and sell it as an over-the-road transportation fuel. This would be a whole new business model and would require the best possible outcomes for our testing, but it is a possible long-term opportunity.

This brings in a second level of regulation and standards to meet. Use of synthetic diesel as a carrier for chemical treatment in the manufacturing process does not require EPA oversight, but this is not the case when it comes to use for road transportation. The EPA has standards regulating what can be used for road transportation, particularly with regard to purity or freedom from contaminants. It would likely be much more difficult for us to use fuel made from CCA-treated wood in this capacity, but chemists tell us that fuel from penta- and creosote-treated wood could very likely qualify for road transportation. Even what we have today could qualify for use as a marine fuel.

We would still have a lot of work to do to get used utility poles approved as feedstock for road transportation fuel, though, as this is heavily regulated. In particular, the EPA recognizes certain classes of waste, such as food, agricultural, and so on, and getting treated wood approved would require creating a new classification or justifying its inclusion under one of the old ones (likely municipal solid waste—i.e., garbage). We would need this type of designation in order to be in compliance with the law to sell our synthetic diesel product as an alternative transportation fuel to traditional diesel, as well as to draw the attention of any of the consumers of this type of product (major petroleum companies that need to have a certain percentage of their inventory, as it were, come from renewable resources or processes).

We also haven't designed our plant around the potential scale of providing a fuel of this kind, which would require a radical change of business model. We would be jumping into a whole new market. This just goes to show, however, the disruptive and transformative results of the type of innovation process we put in place at Cox.

THE LONG VIEW

This unique opportunity, after all, is the product of the sequential innovation process I discuss in the very first chapter. The leadership at Cox deserves a great deal of credit for their commitment to this process and their willingness to take a very long term view, because up front, it is an expensive proposition for us to get involved in this kind of work. They aren't looking for a one-year return on their investment. They realize it may take a few years before we break even, but the potential here is huge.

The potential, which we are very hopeful about fulfilling, is that we will soon be able to go to market and say that we have found a completely circular, patentable, total life cycle solution for how utilities can deal with their waste stream. From a broader perspective, we hope to provide an exemplar of how companies should be thinking about their environmental footprint. All business leaders could benefit from asking themselves about the environmental value they get from their product in the way it is developed, manufactured, and disposed of. In fact, given the new and increasingly circular economy that is developing all around us, business leaders can't afford not to start implementing innovations in a directed way that aims to derive greater environmental value from products and production processes.

CONCLUSION

Business leaders across industries face the challenges of a changing economy and environmental sustainability, and in the face of these challenges, across industries, Cox's potential solutions and productive innovations should be of interest. Cox can serve as a model for business leaders who are interested in implementing a culture of innovation in their own organizations, and in integrating innovations into their overall business strategy.

Cox's innovation process is unique; the same can be said for the specific difficulties and pain points faced by utilities that became the driving force of innovation at Cox. The story told here, though, reflects general themes that business owners should be able to relate to.

Often leaders of small to mid-sized businesses, or of businesses in commodity markets, don't believe that going beyond the status quo is worth the risk. Radical changes in product or business model often seem too costly up front and too unlikely to yield a long-term gain. The Cox example, however, illustrates the importance of long-term thinking about trends in your industry and having a strategic approach to innovating in light of these developments.

In commodity industries, innovation can serve as a brand differentiator and a market disruptor. Companies in these markets don't have to participate in the race to the bottom on prices; positioning yourself as having the most cutting-edge business model in your sector can attract and attach buyers to you even more effectively.

In our case, we are actually putting forth the effort to patent the business model described in this book: recovering used poles, converting the waste wood into fuel, and then circling back around to use that fuel in the production process, thus literally fueling sustainable business. Having this kind of patentable intellectual property can serve as a powerful brand differentiator for your business, and limit the possibility of copycat competitive responses.

The benefits of innovation efforts can also extend beyond the parties involved in the business transaction. In implementing our innovation model, Cox became more environmentally sensitive, at the same time providing a value proposition to end users (the utilities) and the environment and general citizenry.

Companies interested in innovation need to start by examining the likely impact that cultural, economic, demographic, and industry-related issues and trends will have on their business model. In that way, they integrate their innovation efforts into the broader business strategy. Success in this area does not come from brainstorming big ideas in a standalone department whose goals are detached from those of the rest of the company. Success also doesn't come from implementing measures in a reactive manner to meet specific demands here and now. Innovation efforts need to be directed at the long-term aims of the company and reflect long-term trends in the industry.

Rather than rely on luck and happenstance for innovations, companies need to implement a defined process that will generate innovations reliably on a sustained basis, in order to accumulate a company innovation portfolio. Such a process, like the one Cox implemented, is crucial to long-term success. The process must accommodate building a business case for each proposed pursuit, establishing it as a goal that is of a piece with overall company strategy and that everyone in the company can get behind and get on board with.

Those responsible for the innovation process need to have a strong understanding of the company's ability to execute or implement in the various areas of interest. This includes whether the company has the skills and resources needed in-house or will have to acquire them from outside or partner with third parties. If resources for R&D are lacking, companies will need to limit their focus to a narrow spectrum of innovation areas within their portfolios, to avoid spreading themselves too thin and extending beyond their core competencies.

Business leaders need to support these innovation efforts so that the efforts succeed and their companies survive into the future. It is also important to bring a diverse set of stakeholders into the process, including employees with a forward-looking view on the business, customers, and other potential influencers (such as the EPA and other regulatory agencies in our case), and to foster supportive understanding of the value of innovation among the company's leaders and employees.

Finally, as innovation evolves, leaders need to keep an open mind regarding what the endgame may end up being. At Cox, our initial focus was on pole disposal via the currently existing routes, such as waste-to-energy and biomass, but along the way we were exposed to technologies that made us realize we had an opportunity to create a closed-loop manufacturing process that reduced internal costs while promoting environmental responsibility. This involved a radical change in approach and is potentially transformative of our business model, but we believe the potential gain is worth the up-front cost and risk. As is true for businesses across industries, the cost of not innovating in the face of changes in the market and the economy will become too high to sustain in the near future.

OUR SERVICES

To learn more about Barry and his current conversations with electric utilities on the topic of wood waste disposal, visit barrybreede.com. For utilities interested in developing and implementing their own wood waste disposal program, please visit coxrecovery.com for details and assistance.

www.ingramcontent.com/pod-product-compliance
Lightning Source LLC
Jackson TN
JSHW011950131224
75386JS00042B/1653

* 9 7 8 1 5 9 9 3 2 8 9 3 5 *

GETTING OUT OF THE WOODS

One Forest Products Company's Approach to Sustainability and Innovation

This book is about one company, Cox Industries, and their mission to establish an innovation program that powers the creation of a unique, industry-leading business model. Family owned and operated for over sixty-five years, Cox Industries—recently acquired by Koppers—is the largest American-owned provider of treated wood utility poles in the US. With the addition of Cox Industries, Koppers is poised to become one of the largest wood treatment, production, and supply companies in the world.

This book frames the challenges facing many small and mid-sized companies trapped in price-driven, commodity-like industries who remain hopeful of finding a more promising future. In *Transforming the Utility Pole*, author Barry Breede shows how one company and its strategic approach to innovations that are both profitable and progressive can serve as inspiration to business leaders across industries who want to make sustainability and corporate responsibility work together while running a successful business.

BARRY BREEDE is the chief innovation and marketing officer at Koppers Utility & Industrial Products—a national leader in the sale of wood utility poles—leading the company's efforts in commercializing new business ventures, products, and services. Barry also assists Cox Recovery, a Koppers subsidiary providing utilities with environmentally-friendly methods of disposing of wood waste. A graduate of the University of Oregon, Barry has also worked extensively in the innovation area with several global companies including Electrolux AB, Umbro International, and Specialized Bicycles. Barry currently resides in Greenville, South Carolina.

advantagefamily.com

ISBN 978-1-59932-893-5

A LIFETIME OF SENSATIONAL SMILES

TRANSFORMING YOUR CHILD'S LIFE THROUGH ORTHODONTICS

DR. KERRY WHITE BROWN